Dr Agron brings more than 15 years of practical experience in areas of Organizational Leadership and Training Development. He has worked with numerous organizations across countries and industries in research projects, as researcher, trainer and consultant for executives and non-executives. Concurrently, Dr Agron has been teaching in various prestigious top Universities in Malaysia, Turkey, Qatar and Kosovo. Some of courses he taught are organizational behavior, management, industrial and organizational psychology, conflict management, professional orientation, career development and many others.

Agron Hoxha

SUCCESSFUL AGING: LEARN TO GROW OLDER GRACEFULLY

AUSTIN MACAULEY PUBLISHERS™
LONDON • CAMBRIDGE • NEW YORK • SHARJAH

Copyright © Agron Hoxha 2024

The right of Agron Hoxha to be identified as author of this work has been asserted by the author in accordance with Federal Law No. (7) of UAE, Year 2002, Concerning Copyrights and Neighboring Rights.

All rights reserved. No part of this publication may be reproduced, stored in a retrieval system, or transmitted in any form or by any means, electronic, mechanical, photocopying, recording, or otherwise, without the prior permission of the publishers.

Any person who commits any unauthorized act in relation to this publication may be liable to legal prosecution and civil claims for damages.

The age group that matches the content of the books has been classified according to the age classification system issued by the Ministry of Culture and Youth.

ISBN – 9789948759744 – (Paperback)
ISBN – 9789948759713 – (E-Book)

Application Number: MC-10-01-5029194
Age Classification: E

First Published 2024
AUSTIN MACAULEY PUBLISHERS FZE
Sharjah Publishing City
P.O Box [519201]
Sharjah, UAE
www.austinmacauley.ae
+971 655 95 202

Table of Content

Preface	7
Introduction	8
Conceptual Framework	14
Adjustment To Retirement	18
Life Satisfaction	18
Psychological Well-being	20
Stress	23
Depression	26
Factors That Influence Successful Aging	31
Personality	31
Social Network	40
Socio-Emotional Selectivity	42
Family and Children	47
Respect	50
Health	54
Independence	63

Religiosity	*66*
Age	*69*
Education and Income	*73*
Marital Status	*77*
Gender	*81*
Theories: Aging And Adjustment to Retirement	**84**
Social Comparison	*91*
Social Integration	*93*
Psychosocial Resources	*96*
Strongest Factors	**100**
Conclusion	**108**
References	**111**

Preface

This book aims to raise awareness among elderly individuals who often perceive aging as a fixed outcome, rather than an ongoing process. It underscores the significance of recognizing aging as a universal journey of change and personal growth. Adopting this perspective tends to yield a more positive aging experience. In contrast, viewing aging solely as an endpoint can lead to a negative outlook on declining health and well-being. Additionally, this book is pertinent for younger adults who aspire to experience a fulfilling aging process, particularly during late adulthood. Numerous studies highlight that decisions and choices made in young adulthood significantly influence the quality of later life. The primary goal of this book is to clarify that aging itself does not inherently result in negative experiences in late life. Instead, other factors play a pivotal role in determining the quality of life during retirement or late adulthood. The book emphasizes the importance of preparing for successful aging while young, leveraging the social nature of human beings, and cultivating meaningful connections with others.

Introduction

"Age is simply the **number of years** you
have been enjoying."

The World Health Organization (WHO, 2017) has reported that between the years 2015 and 2050, the proportion of the world's older adults is estimated to almost double from about 12% to 22%. This equates to an increase from 900 million to 2 billion people over the age of 60. Current demographic trends suggest that the generations of the world population are aging more than previous ones, with analyses showing that by 2050, older adults (aged 60 or over) will comprise about 20% of the population (Harper, 2014). This demographic shift poses significant socio-economic and health challenges to societies. To address these challenges, it is crucial to study and understand the needs of older adults and evaluate what affects their quality of life. Aging is a universal phenomenon, but it does not occur uniformly.

Adjustment to retirement is a developmental phenomenon that must be understood within the context of lifelong development. Notably, aging or getting older should not connote decline, as decline may refer to unsuccessful rather than successful aging. Individuals should perceive aging as maturation, meaning becoming more patient, tolerant, and accepting of lifestyle changes.

The purpose of this book is to encourage both young and elderly individuals to focus on the importance of aging gracefully rather than merely living a longer life. Longevity or a longer life expectancy does not always imply successful aging, and not every survivor into old age can be considered a successful ager. Successful aging involves focusing on what is important to you and being able to do what you want in old age.

Generally, older people are often described as being depressed, more stressed, less satisfied, and lower in psychological well-being (Hoxha, 2019). However, little research has been conducted to determine the factors influencing their levels of depression, stress, psychological well-being, and life satisfaction. Some elderly people think that the transition to retirement is easy and requires little planning. However, for others, retirement is a significant life change that, despite the freedom and excitement it offers, requires personal adjustment to a new life stage (Barrow, 1996). Quality of life in late adulthood is thus affected by one's lifestyle as a younger adult, and preparation for successful aging should begin in young adulthood. Considering the importance of early preparation for successful aging in determining quality of life and other health-related factors, relevant education and awareness may play a crucial role from young adulthood or even childhood. It is important to investigate the recognition process of healthy aging and identify influential factors in successful aging at each life stage (Jang and Choe, 2006).

This book aims to investigate the factors and reasons why people have different views about later life or adjustment to retirement. Proposed factors that could lead elderly people to have different adjustments to retirement include personality, health, social resources, religiosity, and socioeconomic status. People who reported fewer or no symptoms of depression, stress, high psychological well-being, and good life satisfaction were considered to have good adjustment to retirement (Van Solinge, 2006). Adjustment to retirement refers to the process of getting used to retirement as a new stage of life (Van Solinge, 2006).

In the present study, adjustment to retirement refers to life after retirement, termed as later life or older age. It is often said that people get sick and die shortly after retirement, but studies do not support this notion (Barrow, 1996). Retirement itself does not inherently lead to depression, stress, low psychological well-being, or diminished life satisfaction. There are additional factors that influence how elderly individuals adapt to retirement. While some individuals transition well and successfully adjust to retirement, others encounter difficulties. One man shared his experience, stating, "I turn 80 next February. It's always a milepost, I suppose, but no big deal. I retired about 75 percent at age 65 and then fully retired at age 70. It was easy, very easy, for me to retire. I like the freedom to do what I want, go where I want. I never understood these people who are restless and unhappy in retirement. It seems to me that they have not got much imagination" (Barrow, 1996: 162).

While individuals like the man mentioned face no challenges in adapting to retirement, others do encounter difficulties. As previously discussed, various **factors** contribute to the problems some elderly people experience during their adjustment to retirement. It is evident that retirement alone is not the sole determinant of how individuals adapt to this new phase of life. One of the key factors significantly impacting the process of adapting to retirement is **financial stability**. The level of financial income is closely associated with one's adjustment to retirement. Specifically, individuals with lower income are particularly affected negatively following retirement. Conversely, having a favorable income during later life stages provides elderly individuals with increased opportunities for greater life satisfaction and happiness, ultimately leading to a more positive adjustment to retirement (Barrow, 1996).

Health is another factor that influences elderly people's adjustment to retirement. People in good health are more able to do what they want and go where they want, providing them with the freedom of choice. Those who have freedom of choice, necessitating financial security and good health, reported better life satisfaction and happiness than individuals unable to do what they actually wanted (Barrow, 1996).

Social support was also found to influence adjustment to retirement. People with more friends reported more enjoyable lives compared to those with fewer or no friends. Therefore, they were happier and more satisfied, indicating better adjustment to later life than less sociable individuals. Family is an essential source of emotional support for elderly people. Having family members, close relatives, and friends available when needed is the most crucial support elderly individuals can have (Barrow, 1996). Based on this, social resources are very important in adjustment to retirement. However, having good health alone is not sufficient for successful aging; many interrelated factors contribute to aging success.

People who reported negative attitudes towards retirement or later life appeared to be more introverted in **personality,** tending to be shy, lonely, and lacking interpersonal skills. In contrast, more sociable people with better interpersonal skills, or more extroverted personalities, reported positive attitudes towards retirement or later life (Atchley, 2000). Thus, the way one adjusts to retirement is largely influenced by one's perception of aging and retirement. Therefore, personality type is another contributing factor to retirement adjustment.

Religiosity was found to be an important factor influencing retirement adjustment. Religious people had better skills to cope with aging. Religiosity helped elderly people cope with stressful events, was associated with better life satisfaction, and contributed to better psychological well-being and less stress compared to non-religious individuals (Krause, 2002).

Socioeconomic status played a major role in determining elderly individuals' attitudes towards aging and retirement. The types of work people had before retirement influenced their retirement adjustment. Retiring from non-professional jobs led to happiness, while retiring from professional jobs led to unhappiness (Atchley, 2000). People in professional working-class categories, such as managers, businesspersons, or executives, reported more negative attitudes towards retirement, feeling like they were losing power and social status. Conversely, manual workers like farmers or factory workers tended to have positive retirement perceptions, looking forward to leaving their jobs. However, elderly people with low income and education reported more trouble adjusting to retirement (Atchley, 2000; Barrow, 1996). These mixed results depend on various factors, which will be discussed in the following chapters. Studies confirm that successful or unsuccessful aging from age 70–80 can be predicted by factors experienced before the age of 50 (George, Vaillant, and Keneth, 2001). Therefore, one of the book's main purposes is to explore significant factors to consider before reaching retirement age.

In conclusion, factors indicating elderly people's retirement adjustment included symptoms of depression, stress, psychological well-being, and life satisfaction. The main factors influencing retirement adjustment were personality type (introverted, extroverted, etc.), health (independence, mental and physical health), social resources (respect, family, and social network), religiosity, and socioeconomic status. This book aims to uncover the factors contributing to the support requirements and influencing the quality of life of elderly individuals with good physical and mental functioning. Additionally, the book explores the relationships and interconnections among these factors.

Conceptual Framework

This book investigates the impact of independent factors such as personality (perception of aging), health (independence, mental and physical health), social resources (respect, family, children, and social network), socioeconomic status (age, education, income, and marital status), and religiosity on dependent factors including depression, stress, life satisfaction, and psychological well-being as depicted in Figure 1 below.

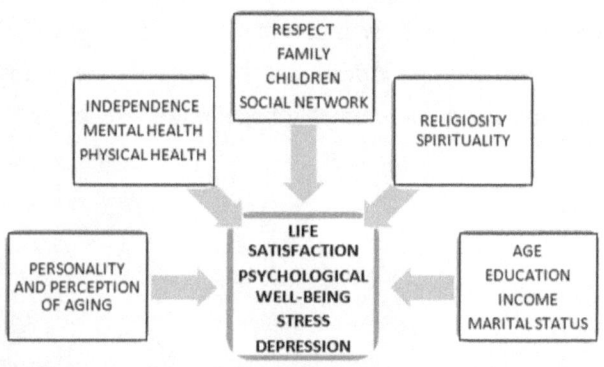

Figure 1: Factors Influencing Adjustment to Retirement

In this book, depression refers to the mood disorder characterized by a state of sadness, gloom, and pessimistic ideation with a loss of interest and pleasure in normally enjoyable activities. Stress refers to emotional or mental strains caused by the difficulties of daily life. Psychological well-being refers to how elderly people feel about themselves in later life or about old age. Life satisfaction refers to the quality of life in later life; it concerns how satisfied elderly individuals are with their life after retirement.

The following parts of the book will discuss the common complaints of psychosocial factors among elderly people, including life satisfaction, psychological well-being, stress, and depression in later life. The author has selected these four main complaints based on 170 interviews conducted with elderly people aged 65 and above. These factors have been further investigated using relevant standardized instruments. The next goal was to investigate and identify the main determinants of life satisfaction, psychological well-being, stress, and depression.

The main goals of the study were:

- To analyze the relationship between personality and the level of depression, stress, psychological well-being, and life satisfaction among elderly individuals in later life.
- To analyze the relationship between health (mental health, physical health, and independence) and depression, stress, psychological well-being, and life satisfaction among elderly people.

- To analyze the relationship between social resources and depression, stress, psychological well-being, and life satisfaction among elderly people.
- To analyze the relationship between religiosity and depression, stress, psychological well-being, and life satisfaction among the elderly.
- To analyze the relationships between different levels of socioeconomic status (age, education, income, marital status) and the level of depression, stress, psychological well-being, and life satisfaction among the elderly.

In this research, standardized instruments were used to analyze the participants' personalities and determine which personality types exhibited a more positive outlook on aging after retirement. Through interviews, it was observed that individuals with poorer health tended to have more negative experiences of aging in later life compared to those in good health. Furthermore, individuals who had larger social networks, more family members, and who were well-respected by others reported a more positive perception of aging after retirement, indicating a higher level of successful aging. Religion also emerged as a significant factor in coping with aging, particularly during late adulthood.

Additionally, this study incorporated the findings of multiple scientific studies that explored similar topics. The results and recommendations from these collective studies were supported by prominent theories such as Erik Erikson's lifespan stages, social comparison, social integration, and psychosocial resources. These theoretical frameworks provided further insights into the questions addressed by the research and strengthened the overall conclusions.

Adjustment To Retirement

"Age is like climbing a mountain; the higher you climb, the better view you get."

This part of the book presents studies related to depression, stress, psychological well-being, and life satisfaction. It also reviews factors such as personality, mental health, physical health, independence, respect, family and children, social network, religiosity, and socioeconomic status (age, gender, education, and income) in association with levels of depression, stress, psychological well-being, and life satisfaction. Theories of aging and retirement will be discussed to show how these factors are associated with depression, stress, psychological well-being, and life satisfaction.

Life Satisfaction

Life satisfaction is a pivotal criterion for describing the aging process. Higher levels of life satisfaction in later life are indicative of a higher quality of the aging process. Furthermore, previous researchers have not differentiated between life satisfaction and quality of life (George and

Bearon, 1980). It has been reported that happy elders have higher levels of life satisfaction and quality of life. Life satisfaction can be defined as a sense of satisfaction or pleasure regarding one's present and past life. A perspective on successful aging accepted by social gerontologists is the life satisfaction approach, which posits that individuals have aged successfully if they feel happy and satisfied with their current and past endeavors. Rudinger and Thomae (1990) assert that the focus of subjective well-being has been on explaining how people experience their lives, including their cognitive assessment, emotional reactions, and adjustments to later life. Life satisfaction is generally considered a cognitive aspect of subjective well-being, implying a judgmental process where elderly people compare their current life situation to their desired life satisfaction (Diener et al., 1999; Krause, 2004). These findings suggest that it is the cognitive perception of elderly individuals about their later life and their expectations at this stage. The challenge for elderly individuals lies in matching their life expectations with their desired outcomes, emphasizing the need for people to develop a realistic and compassionate understanding of what to expect or desire in later life. Additionally, to enhance life satisfaction in later years, proactive preparation is essential. This preparation enables individuals to better navigate the transitions and changes associated with aging and maximize their later life stages. Retirement can provide an opportunity for further self-development and achieving life satisfaction. Later in this book, factors affecting life satisfaction will be discussed in detail. For instance, personal health is a major factor determining the level of life satisfaction. This factor becomes more difficult to control if proper health care is not taken before reaching retirement or the later stages of life. In this context, Jik-Joen-Lee (2005) examined 109 individuals aged 60 and over who were living alone in public housing estates in Hong Kong, although they had children. The study aimed to identify factors associated with life satisfaction and found that, among many factors, health predicted the quality of life. Conversely, poor health can worsen life quality by increasing medical care costs, reducing social interactions, and limiting the ability to visit preferred places or participate in various activities. This example illustrates how health can significantly influence experiences in later life. Therefore, it is crucial to take care of one's health before it is too late and to cease habits that deteriorate health while still healthy. Another determinant of elder satisfaction depends on personal goals. Thus, it is highly recommended that elderly individuals plan and set personal goals for growth rather than disengage from the outside world. Research has shown that individuals lacking personal goals during later adulthood tend to have lower life satisfaction. This book will explore many other factors that can aid in preparing for an improved later life.

Psychological Well-being

Psychological well-being is a complex term that often confuses people about its meaning. Generally speaking, well-being refers to feeling physically and mentally healthy and having a good life (Kibret and Tareke, 2017). The well-being of elderly people depends on how they perceive their experiences, including feelings of happiness, sadness, anxiety, or excitement (Tinkler & Hicks, 2011). Many studies have focused on the correlation between psychological well-being and older age. Psychological well-being is a variable with different meanings for different social scientists. However, a commonly accepted definition would be what researchers often term subjective well-being. Subjective refers to personal evaluation based on how the person feels about him/herself. Therefore, combining the definitions of subjective and well-being refers to a person who feels physically and mentally healthy, based on their own assessment, rather than on others' diagnoses. In addition, we can say that subjective well-being and psychological well-being are synonymous for the most part (Barrow, 1996). It is safe to assert that improving older adults' well-being can potentially contribute to successful aging.

Although psychological well-being is important for all age groups, it is particularly relevant to older people for several reasons. It is estimated that in 20 years, nearly a quarter of the UK's population will be aged 65 and over (Medical Research Council, 2010). As life expectancy increases, there is a growing emphasis on preserving the psychological well-being and morale of older individuals. Consequently, it is increasingly vital to implement preventive measures rather than focusing solely on maintaining psychological well-being in later life. Understanding the factors that impact the quality of psychological or subjective well-being is crucial in this context.

What influences psychological well-being is a significant concern. For the last 60 years, scientists have sought to determine what decreases psychological well-being in later life. Cagney and Lauderdale (2002) found that elderly people with higher levels of education, income, and wealth reported higher psychological well-being compared to those with lower levels of education, income, and wealth. Additionally, psychological well-being was found to be associated with spiritual beliefs and health. Individuals with weaker mental and physical health exhibited lower levels of psychological well-being. Furthermore, elderly people involved in religious activities reported higher levels of psychological well-being than those not involved in religious activities (Edison, Boardman, William, and Jackson, 2001; Kirby, Coleman, and Daley, 2004). There are factors such as religiosity, social activities, and exercise that elderly people can improve in later life, but there are other factors to consider in advance, such as education, health, and other factors crucial to well-being. Studies have found that psychological well-being is lower at older ages compared to younger generations, but other studies contradict this (Zaninotto, Falaschetti, and Sacker, 2009). It is not age itself that leads to lower quality of psychological well-being; other factors influence this quality. Controlling for health, socioeconomic status, and psychological factors, there are no overall cohort differences in quality of life between older and younger adults. It is not appropriate to assume that quality of life inevitably diminishes with age. Over time, factors such as depression, functional limitations, limited wealth, lack of positive support from spouse, children, and friends, and having a small social network have been found to detrimentally affect the quality of life. These factors can impact younger generations, but younger individuals have more time to address and enhance these factors in their lives. Therefore, it is suggested that for better quality of aging, people must be aware of the factors that have a stronger impact on the quality of psychological well-being than aging itself. The general finding of a large body of gerontological literature is that there is no age-related decline in well-being; rather, it is the aforementioned factors that influence well-being and quality of life.

Stress

The level of stress was reported to be an important element influencing adjustment to later life, with social resources suggested as essential for reducing stress levels and aiding in coping with a stressful environment. Further, the key factor to controlling stress in later life was found to be staying healthy (Hayer et al., 1999). Hawkley and his colleagues conducted a study examining the conceptualization of stress and resistance to life stress. They reported that individuals experiencing severe stress, usually due to employment problems or interpersonal difficulties, were at a higher risk of developing diseases (Hawkley, Berntson, Engeland, Marucha, Masi, and Cacioppo, 2005). Psychological stress can trigger adverse behaviors, leading individuals to engage in detrimental habits such as smoking, improper eating, insufficient sleep, and lack of exercise. These negative practices can contribute to physiological ailments, including cardiovascular disease or a weakened immune system. However, individuals who manage and regulate their behaviors are more likely to reap the benefits of good health. Physical exercise and muscular development were found to be crucial in enabling the body to respond effectively to stress. Hawkley et al. (2005) exemplified this by presenting the case of a student pianist who transitioned from graduate school to professional performances, requiring more practice time. If the pianist had well-developed muscles in the arms, back, and hands, they would likely experience less fatigue and stress compared to someone whose muscles in these areas were not as developed. Additionally, in this study, lonely and non-lonely individuals were interviewed to determine which group reported higher levels of stress. Lonely individuals reported more stressful times than non-lonely individuals (Hawkley, Berntson, Engeland, Marucha, Masi, and Cacioppo, 2005). As mentioned in other studies, prolonged levels of stress can lead to more significant problems, including depression. Another finding from this study was the difference in perceived stress between older and younger adults. Perceived stress among older adults was found to be strongly associated with higher levels of Systolic Blood Pressure (SBP). It was concluded that although stress levels increase with age, the possibility exists to minimize stress if individuals maintain good health and resilient physiology (Hawkley, Berntson, Engeland, Marucha, Masi, and Cacioppo, 2005). Research has shown that as individuals age, they become more capable of managing stress, having developed successful strategies through experiencing many stressful life events. However, older adults are more exposed to stress due to the increased frequency of stressful life events, such as the death of a spouse, friend, or close relatives, which are more likely to occur in later life than during early adulthood (Hayer et al., 1999). Martin, Grunendahl, and Martin (2001) examined 489 participants aged 41 to 43 and 449 older adults aged 61 to 63. The study compared the availability of social resources, stress, and well-being between these two age groups. Stress can be generated from several factors, such as declines in health, social network size, and economic status, which tend to decrease with age. This book primarily focuses on the relationship between social resources, stress, and well-being. When there is an increased level of stress and a decrease in individual resources, social resources can compensate for individual loss and help reduce stress levels in later life. Health was found to be the most influential factor affecting stress levels in later life, as physical functioning, which determines the well-being of older individuals, becomes limited. Health affects older adults because as they age, their stress levels increase while their coping resources diminish. In other words, at the time when financial and social resources are most needed to cope with aging, they are also decreasing. It was concluded that social resources were negatively associated with levels of stress; the higher the social resources in later life, the lower the stress levels. The lower the stress levels, the higher the level of well-being in later life (Martin, Grunendahl, and Martin, 2001). In addition, Krause (2003) reported that stress is usually associated with a decline in health and well-being. Moos and his colleagues (2005) made significant findings regarding the impact of social interactions within networks such as children, spouses, relatives, and close friends on stress levels. They found that conflicts or criticisms within these relationships could contribute to increased stress levels and symptoms of depression. The study also suggested that nurturing and maintaining positive relationships in later life could help older individuals effectively cope with stressful events, as heightened stress levels have the potential to impact the onset of depression. Furthermore, the study highlighted that factors beyond interpersonal relationships, such as health and financial status, as well as conflicts within relationships, could generate stress. It was observed that individuals experiencing higher levels of stress were more prone to developing symptoms of depression in later life. Therefore, an association was identified between interpersonal conflict, stress, and depression (Moos, Schutte, Brennan, and Moos, 2005).

Depression

Both elderly people and caregivers must understand that depression is not a normal part of aging. The lack of effective communication and interaction between mental health and medical professionals has resulted in increased cases of depression among elderly people.

The National Institute of Health (NIH) Consensus statement on the diagnosis and treatment of depression in later life (1992) reported:

> "Many senior citizens will live their final years in despair and suffering without any appreciation of their affliction or the understanding and comfort from those most dear to them. Professional help is not often sought or offered, and depression is not likely to be brief. The likely consequences are loss of personal happiness and severe strain on living circumstances. Depression may trigger a shift from home to a nursing facility, or may shift from a warm and respected friend or loved one to an isolated individual with lost status. Untreated depression costs money because physical illnesses require more medical services, living arrangements become institutional, and employment is lost. These costs should be substantially preventable with presently validated

case recognition and treatment techniques" (Katz et al., 1998: 120).

Older individuals are seemingly more affected by depression compared to younger people. Research estimates suggest that around 10-15 percent of individuals aged 65 and above experience depressive symptoms (Kraaij, Arensman, and Spinhoven, 2002; Turner, Lloyd, and Taylor, 2006). Other studies indicate that the percentage of elderly individuals with depressive symptoms may be even higher. Depression is commonly associated with older people, but it is important to note that psychological conditions like depression or anxiety are not a typical aspect of the aging process (Everard et al., 2000).

Recent studies have reported that younger adults are as vulnerable as older adults to experiences of depression. Although depression is a common problem among older adults, it is not a normal part of aging. In fact, studies show that most older adults feel satisfied with their lives, despite having more illnesses or physical problems. However, there are some factors that cause elderly people to experience more symptoms of depression. For instance, the death of a spouse, financial problems, and poor health were identified as potential contributors to depressive symptoms as well (Hayer, Rybash, and Roodin, 1999). After a period of adjustment, many older adults can regain their emotional balance, but others do not and may develop depression (Zarit, Todd, and Zarit, 1986).

According to recent studies, there is a concerning issue among the elderly population where many individuals associate their later years with feelings of unhappiness, which can contribute to the development of depressive symptoms. Regrettably, a notable number of older adults lack awareness regarding the symptoms of depression or exhibit hesitation in seeking the necessary support to address their condition.

For instance, scholars have suggested that certain issues such as decreased mood, disrupted sleep, decreased appetite, loss of interest, anxiety, and reduced energy serve as indicators of depression among older adults. However, a significant proportion of elderly individuals who experience depression do not pursue treatment due to difficulties in recognizing their condition. They may perceive feeling down as a normal part of life or believe that these problems are inherent to the aging process. Moreover, elderly individuals tend to resist leaving their homes and prefer isolation, which is also a sign of depression that they may not be aware of. Additionally, they are often reluctant to discuss their emotions or seek assistance.

Failure to recognize the symptoms of depression and not taking necessary steps for treatment can result in other serious health problems. It was reported that depressed people are more vulnerable to distress, disability, morbidity, cognitive impairment, and to an extent, mortality (Katz et al., 1998). However, many people mistakenly believe that these issues are a natural part of aging when, in reality, they can be prevented with proper intervention.

Everard et al. (2000) conducted an extensive study to investigate the prevalence of mental disorders in older adults. The researchers aimed to determine why some elderly individuals develop depression while others do not. Engaging in low-demand leisure activities was found to be linked to improved mental health. This finding aligns with recent studies that have also identified a connection between less physically demanding activities and better overall functioning. These actions are easily noticeable, and it's important to inform older adults about the importance of changes in their activities. If older adults or others notice a decline in their regular activities or a shift from high-demand leisure activities to low-demand ones, steps can be taken to address the situation before it leads to limitations in functioning. Intervening during the early stages may have a higher chance of success in reducing disability and, ultimately, healthcare expenses compared to interventions done at a later stage. A study conducted in Japan revealed that individuals with physical disorders and elderly individuals over the age of 65 who had survived a stroke displayed higher levels of depressive symptoms compared to those without physical disorders. Therefore, the book suggests that taking care of physical health is crucial in reducing or even preventing symptoms of depression. The interrelation between depression and physical health is evident, as studies have reported that individuals confined to bed showed 100% more depressive symptoms than those who were not bedridden. In older adults, depression is typically linked to other physical disorders (Shinkawa, Yamaya, Ohriu, Hiroyuki, and Sasaki, 2002; Everard et al., 2000). These findings indicate that older people are more vulnerable to a decline in physical functioning, independence, quality of life, and perceived well-being. It was concluded that there is a strong association between psychological disorders and physical health disorders among older individuals. Consequently, being mentally unhealthy in later life, in combination with physical disorders, increases the risk of experiencing major symptoms of depression and anxiety (Everard et al., 2000).

A comprehensive study conducted by Kraaij et al. (2002) aimed to explain the relationship between various negative life events and depression in old age. The study established a significant association between negative life events and depression, identifying traumatic events such as the death of a family member or close friend, daily hassles, witnessing fatal or severely injurious incidents, personal harm or harm to close associates, severe illness, and low socioeconomic status as factors that could contribute to depression.

Furthermore, Diwan, Jonnalagadda, and Balaswamy (2004) discovered a significant association between depression and poor health, slower recovery from illnesses, and lower self-rated health. Involvement in religious activities was linked to fewer symptoms of depression, loneliness, and anxiety. Based on the aforementioned studies, it can be concluded that individuals aged 60 and older tend to experience more depressive symptoms compared to earlier stages of life. Factors such as a decline in physical functioning, limited social networks, remaining life expectancy, and financial security may contribute to depression in later life. These studies suggest that elderly individuals can potentially have a more positive perception of aging by managing factors such as health, social networks, and financial security. Instead of expecting a decline in health, social networks, and financial security, individuals should focus on maintaining these aspects and avoiding factors that contribute to signs of depression. The next chapter of the book will explore the psychosocial factors that play significant roles in successful aging.

Factors That Influence Successful Aging

"Be happy for this moment. This moment is your life." –
Omar Khayyam

The term "successful aging" was made popular in 1987 when scientists John Wallis Rowe and Robert Kahn published an influential book entitled "Successful Aging." According to Rowe and Kahn, successful aging is achieved when individuals are free of disability, have cognitive and physical abilities, and are socially engaged in desired ways. In other words, successful aging is considered when individuals are able to continue with their daily activities as before, physically healthy to handle their daily activities, and mentally healthy for understanding and analyzing others. The following section explains the role of personality in successful aging.

Personality

Personality traits are defined as consistent and enduring dispositions involving individual differences in the style or content of thought or behavior. Personality has been noted to play an important role across all stages of life. It is widely accepted as one of the strongest and most consistent predictors of well-being and life satisfaction for elderly people. Researchers have suggested that different personality traits may cause cognitive changes in older adults. Hence, personality is undeniably a very important aspect of successful aging. There are five types of personality traits that have a direct influence on how individuals perceive aging. These five personality traits are openness, conscientiousness, extraversion, agreeableness, and neuroticism.

Openness refers to people who are open and motivated to learn new things and enjoy new experiences. Openness is highly associated with awareness of inner feelings, a need for variety in activity, intellectual curiosity, and tolerant value systems. People who score high in openness are associated with traits like being insightful and imaginative and having a wide variety of interests. The trait of openness could be a valuable resource for older adults' life satisfaction because it could aid them to face a crucial developmental challenge based on the establishment of a satisfactory post-retirement lifestyle (Van Solinge & Henkens, 2008). Older people that score high in openness to new experiences are more likely to engage in activities that help them maintain their health and level of functioning. Thus, openness to experience is an adaptive personality variable that tends to develop higher levels of self-esteem and life satisfaction (Chan, Luciano, and Lee, 2018). In addition, older adults could be better equipped to cope with age-related physical and social challenges, which could contribute to the maintenance of their life satisfaction (Kling, Ryff, Love, and Essex, 2003; Jurkuvenas, Zamalijeva, Pakalniškienė, and Kairys, 2017). This book encourages post-retirement individuals to consider openness to experience as a valuable resource for life satisfaction. Previous studies have reported that openness to experience has a positive impact on the development and maintenance of a healthy lifestyle during the post-retirement period.

Conscientiousness, on the other hand, explains people with higher levels of reliability. Conscientiousness is a fundamental personality trait that reflects the tendency to be responsible, organized, hard-working, goal-directed, and to adhere to norms and rules. People who score high in conscientiousness are reported to be more organized and detailed than people with lower levels of conscientiousness. Individuals who score high on conscientiousness are more likely to work hard towards their goals, work long hours, and persist after failures or setbacks. Since higher levels of conscientiousness are linked to calm, organized, and responsible behaviors, these individuals are more likely to avoid risky behaviors that may cause physical symptoms or disease. Being aware of the importance of conscientiousness will add value to successful aging. Apart from improvement in physical health, individuals with higher levels of conscientiousness were associated with a slower rate of cognitive decline as well.

Extraversion refers to outgoing people who get their energy from interacting with others, while introverts get their energy from within themselves. These people tend to be more energetic, talkative, and assertive compared to people who score higher in introversion. Extraverts are expected to have sustained levels of life satisfaction as they age (Mroczek & Spiro, 2005). Individuals with high levels of extraversion have a tendency to see events in their life as challenges instead of threats. Positive extraversion might work as a protector against cognitive decline by encouraging older adults to engage in intellectual activities and by facilitating a positive attitude toward aging (Human et al., 2013). These findings indicate that individuals experiencing decreasing levels of extraversion as they age report weaker perceived cognitive functioning, physical functioning, and physical health compared to individuals who maintain their high levels of extraversion.

Agreeableness explains individuals who tend to be friendly, cooperative, and compassionate. People who score low in levels of agreeableness may be more distant and isolated. People who score high in agreeableness tend to be kinder, more affectionate, and more sympathetic than individuals with lower levels of agreeableness. High levels of agreeableness are mostly related to maintaining relationships (Jensen-Campbell & Graziano, 2001). Their motivation to maintain positive relationships encourages them to follow up with necessary treatment plans and activities prescribed by their medical doctors or health consultants. In general, agreeableness is a personality trait that promotes positive social relationships, prosocial behaviors, effective conflict resolution, and a supportive social environment. Embracing agreeableness can contribute to a more fulfilling and successful aging experience. Therefore, individuals must work on maintaining the levels of agreeableness for better aging.

Neuroticism refers to the degree of negative emotion in individuals. Sometimes, neuroticism is also known as emotional stability or instability. Individuals high in neuroticism are likely to see the problems of old age as crises and to complain about poor health. Neuroticism is a powerful determinant of life satisfaction, morale, well-being, unfounded health complaints, and negative-affect scores. Individuals scoring high on neuroticism often experience emotional instability and negative emotions such as moodiness and tension. High levels of neuroticism are related to low life satisfaction (Mroczek and Spiro, 2005), and low subjective well-being (Jurkuvenas, Zamalijeva, Pakalniškienė, and Kairys, 2017). The feelings of negative emotion or depression, such as high neuroticism, might have deleterious effects on brain functioning. Individuals with high scores on neuroticism tend to perceive events around them in a more negative way (Lahey, 2009), and they tend to respond to events and people with negative emotions and frustration. In other words, levels of neuroticism predict the quality of physical and mental health at all age groups, but more significantly at older age groups. Therefore, for young and older adults, it is very important that they are aware of the consequences of neuroticism after retirement. Everyone must realize that as people age some senses weaken and some activities slow down, but our ability to love and be loved does not diminish with age. Elderly individuals, in particular, must focus on the positive sides of aging and accept aging as change, maturation, and freedom rather than disengagement or isolating themselves from others. Each of these personality types is found to play an important role in experiencing successful aging in terms of physical symptoms, psychological well-being, and social interaction. For instance, a high level of neuroticism was found to be a critical factor in predicting the quality of psychological well-being (Luchetti, Terracciano, Stephan, & Sutin, 2016). These researchers found that individuals with high levels of neuroticism or low emotional stability are more vulnerable to experiencing stress and difficulty in dealing with stressors. Extroverts, on the other hand, tend to experience more positive emotions and social support that may help them remain physically and mentally healthy. Similarly, individuals with high conscientiousness are careful and systematic in decision-making, which helps them avoid engaging in harmful behaviors. Individuals high in agreeableness are also reported to be helpful, cooperative, and socially interactive.

Openness, conscientiousness, extraversion, agreeableness, and neuroticism were found to significantly impact subjective well-being (Chan, Luciano, and Lee, 2018). For example, agreeable individuals tend to be more skillful at obtaining social support, but neurotic individuals are more likely to be exposed to interpersonal stress. On the other hand, people scoring higher in conscientiousness, extraversion, and openness are more likely to perceive events as challenges rather than threats. Previous studies indicated that low scores on neuroticism and high scores on extraversion, openness to experience, agreeableness, and conscientiousness are significantly helpful for successful aging. Higher levels of emotional stability, agreeableness, and conscientiousness were positively related to higher levels of cognition, a higher likelihood of engaging in volunteer work, higher levels of activities of daily living, and higher levels of subjective health. Extraversion and openness to experience were also positively associated with cognition and engaging in volunteer work (Baek, Martin, Siegler, Davey, and Poon, 2016). In order for elderly people to have better aging quality, they are strongly encouraged to be more aware of the type of personality they have, and what type of personality is necessary for better life satisfaction. Furthermore, awareness of personality in older adults helps adapt to a new environment based on their enduring personality traits, decide how to react and interact in critical situations, and develop self-identity, which are essentials needed to age successfully.

Individuals possess unique and relatively stable personality traits that are not easily altered. This book does not promote or propose the idea of changing one's personality. However, it emphasizes the significance of self-awareness regarding the expected manifestation of specific personality traits in various situations and with different individuals. Recognizing and understanding one's own individual traits is vital, as it enables individuals to comprehend how each trait can influence their overall well-being and life satisfaction.

In some cases, personality types may indirectly influence life satisfaction, such as by affecting health issues and relationship difficulties. Therefore, it is highly important for individuals to understand and accept the various styles of personalities that people can possess. This awareness allows individuals to understand the expectations placed upon them regardless of their actual personality type. For instance, even individuals who naturally lean towards introversion and lack a strong inclination for socializing may find it beneficial to engage socially on certain occasions or with specific individuals. This is because extensive research consistently indicates that individuals with more extroverted tendencies tend to have a more successful aging experience compared to their introverted counterparts.

It is worth noting that feelings of loneliness can impact individuals of all age groups and are not solely determined by the aging process. Some individuals may experience a sense of isolation and a lack of companionship, but these emotions cannot be solely attributed to their age. It is crucial to recognize that aging and loneliness are not inherently intertwined. It is intriguing to note that individual personality traits can impact how aging is perceived. Some individuals have a positive outlook on aging and view it as a natural and normal part of life, even when faced with challenging circumstances. This perspective allows them to navigate the aging process with a sense of acceptance and adaptability. In a comprehensive study conducted by Isaacowitz and Smith in 2003, 516 participants aged 70 and older from Berlin were examined to investigate the factors that predict positive and negative effects in old age. The researchers aimed to understand the variables associated with feelings of happiness, joy, life satisfaction, as well as distress, fear, and anger. This research shed light on the intricate relationship between personality, aging, and emotional well-being. It highlighted that personality traits can play a significant role in how individuals perceive and experience the aging process. Some individuals, despite facing challenging circumstances, possess an optimistic outlook on aging, which contributes to their overall positive emotional state in later life. It is crucial to understand that aging itself is not a direct cause of negative emotions or loneliness. Instead, it is the complex interplay between personality traits, life circumstances, and individual perspectives that shape the emotional experiences of older individuals. By acknowledging and addressing these factors, we can foster a better understanding of aging and develop strategies to promote well-being and alleviate feelings of loneliness among older adults.

Personality was found to be a strong predictor of the aging experience; it was an even stronger predictor than health (Isaacowitz and Smith, 2003). Individuals who scored high in extraversion were reported to have more positive experiences of aging, whereas individuals who scored high in neuroticism were reported to have more negative experiences of aging. Thus, it was concluded that extraversion predicted positive affect, and neuroticism predicted negative affect in later life (Isaacowitz and Smith, 2003). Extroverted adults were reported to be less lonely in later life. They were reported to have larger social networks than adults who were not extroverts. Extroverts were warmer and more sociable. Their social network could provide them with emotional support even after the death of a spouse or if lacking family interaction (Lang and Martin, 2000). Older people who had high self-esteem reported having good control and the ability to maintain their health. It was found that individuals who evaluated themselves as high in self-esteem were less likely to suffer from chronic illnesses (Bailis and Chipperfield, 2002).

Featherstone and Hepworth (1991) suggested two types of aging images which are distinguishable among elders. The first one is called 'heroes of aging,' which refers to elders who have a positive attitude towards aging. They usually take care of themselves and do not act like being old but, on the contrary, try to be young all the time in their daily actions, bodily posture, and general behavior. These people are regarded as optimists in other studies; they perceive aging and stressful events as challenging and not as threats to their life. People like 'heroes of aging' and optimists are more likely to show improvement in psychological well-being and have better physical health. The second image of aging is that some elders experience severe bodily decline through disabling illness to the extent that the outer body is seen as misrepresenting and imprisoning the inner self. This type of image is called 'mask of aging' and shows the more disagreeable aspects of the aging process and a negative attitude towards aging. They perceive aging and the problems that come with aging as threats to their life. These people are more likely to decline in their physical health and life satisfaction (Featherstone and Hepworth, 1991; Uskul and Greenglass, 2005).

Social Network

"Older individuals who adjust to later-life transitions by remaining socially active are happier and healthier."

Two intriguing findings emerge regarding the connection between aging and social networks. One set of researchers observed that individuals in later life tend to experience a disconnection from friends and social groups. However, contrasting findings from other researchers indicate that older adults exhibit a tendency to engage in more social interactions within their neighborhood and participate in religious activities. These two findings indicate that elderly individuals do not really disengage as they age, but they change the direction and form of social connectedness. Traditionally, some classic perspectives described elderly life as a time of loneliness and rolelessness, a perspective that might have negatively influenced the self-perception of elderly individuals. However, more recently, researchers have continuously worked on increasing the awareness of individuals about the importance of staying active and socially engaged in the post-retirement period as well. Prominent researchers of gerontology have repeatedly shown that social integration is the key component of "successful aging" (Rowe and Kahn, 1998). Therefore, as elderly individuals disconnect from some groups and activities, they must connect and integrate into other groups and activities. This way they can manage to stay active and socially engaged, hence have more successful aging. If elderly individuals, on the other hand, tend to distance themselves from some groups and activities, and decide to isolate themselves from society, they are more likely to have lower life satisfaction.

In previous literature, it was reported that the quality of social networks might have a great impact on depression and psychological well-being. Fiori et al. (2006) reported that the results of their study supported previous findings that individuals with small social networks have higher depressive symptoms not only because they lack the companionship of friends but also because they are more socially isolated. Therefore, it is this lack of social integration that negatively influences health (Fiori, Antonucci, and Cortina, 2006). It can be said that psychological well-being would be improved by belonging to a close social network because individuals may consider themselves as being loved and cared for by their network members as compared to those who are socially isolated. However, in certain cultures and countries, older individuals start gradually to distance themselves from social roles, narrowing role sets, and weakening of existing social relationships. These factors such as distancing themselves from social roles, narrowing role sets, and weakening of existing social relationships are the main causes of adults' isolation from social networks. Therefore, it is highly recommended that elderly individuals must continue with maintaining social networks as they age.

It is well known that older adults who adjust to later-life transitions by remaining socially active are happier and healthier. It is also important to understand that social support from social networks during old adulthood is strongly affected by habits that individuals practice before the age of 50 (George, Vaillant, and Keneth, 2001). This study indicates that to have a healthy social network in late adulthood, individuals must think while still young because all studies confirm that a healthy social network leads to more successful aging than a weak social network.

Socio-Emotional Selectivity

Adjustment to retirement can be affected by multiple factors, including a person's psychological, behavioral, and social well-being. Emotional support helps older individuals cope with loneliness and low levels of hope. Without emotional support, a person feels lonely, and loneliness leads to negative experiences of aging (Dykstra, 1995).

Socio-emotional selectivity is an issue not fully explored, but what is known is that the context of social functioning is one of the factors influencing socio-emotional selectivity (Dykstra, 1995). Socio-emotional selectivity refers to people who give up their long-term goals, such as education and ambition. Instead, they shift to short-term goals, such as emotional support and immediate fulfillment of their desires;

as their remaining time to live decreases, they try to move towards emotional relationships. Social context refers to the availability of social context opportunities, which may influence the adaptiveness of certain social patterns.

An increase in social support reduces the risk of physical disease and mental health issues. Social support can be a perceived or real resource provided by others, which is very important for people who are dependent on others, like family, friends, and children. Social support makes the person feel cared for, respected, and valued.

Oxman and Hull (2001) reported a strong association between social support and depression among older people. Social support is often associated with less impairment in activities of daily living (ADL). People who are withdrawn from society show higher levels of stress, depression, and impairment in activities of daily living. The greater the number of close friends having frequent contacts, the fewer symptoms of depression will be present.

The feeling of being needed and respected by others gives older people the strength to take care of themselves. This is a very important motivation for the elderly to be more aware of their physical and mental health. Being supported in later life is very protective because social support gives older people meaningful roles in their lives. Lack of social support and close attachments, on the other hand, generates emotional loneliness and depressiveness, and these people could be at higher risk for death compared to individuals who are socially supported and valued. More frequent contacts among close friends were found to lessen symptoms of depression (Lyyra and Heikkinen, 2006).

To confirm that social support is related to health-related quality of life (HRQOL), the Center for Disease Control (CDC) analyzed data from the 2000 Missouri Older Adults Needs Assessment Survey (MOANAS) of adults aged 60 and older (Keyes, Michalec, Zahran, and Kobau, 2005). The Missouri Department of Health and Senior Services conducted a statewide, stratified, random digit-dialed telephone survey of 3,112 participants. The survey was designed to collect health, social service, and needs assessment data on Missouri adults aged 60 and over. In this study, it was found that visits with friends or relatives, having close friends for emotional support, and the perception of help being available if sick or disabled were associated with better health and particularly with better mental health among older adults.

In this study, Keyes and his colleagues (2005) reported that older adults who almost never visited their friends during the preceding 30 days were more mentally unhealthy and had fewer vital days than those who visited friends or relatives at least several times a week (1.9 vital days versus 5.1 vital days); they were sadder or more depressed (5.7 days versus 2.1 days). People with no close friends were more mentally unhealthy than those with three or more friends (8.1 non-vital days vs. 5.5 non-vital days). In addition to these findings, people who perceived that help was available had fewer unhealthy days than those who perceived that no help was available. The study concluded that those who had social support, either emotional, instrumental, or informal support, were in better physical health and mental health (Keyes, Michalec, Kobau, and Zahran, 2005).

Lang and his colleagues (1998) conducted a study of 516 participants, aged 70 to 104, using intensive 14-session interviews and found that although the social network gets smaller in very old age, the number of close emotional relationships was maintained. It was found that due to the awareness of the elderly individuals about social-emotional selectivity, elderly individuals tend to forget about long-term goals and focus more on short-term goals. In other words, as individuals get older, they shift their preferences to immediate benefits, and therefore are more likely to receive emotional support. By emotional support, the researcher meant focusing on and maintaining close friendships and family. The researcher tried to explore whether the indicator of socio-emotional selectivity was related to personality characteristics and family status, and the results showed that at least two factors had a positive influence on socio-emotional selectivity, namely the social context and personality characteristics of elderly individuals.

Social context, on the one hand, influences both the size and level of emotional closeness in the network. Social context influences the availability of external and social resources such as family and close friends in later life, which would help in adaptiveness. For example, if an adult individual lives with a nuclear family of a wife or children, he/she may get emotional support easily because family members are reported to be essential emotional resources. However, a person with no nuclear family but with close friends in a social network may experience an even greater feeling of social embeddedness because close friends substitute for a nuclear family. It was further found that people with larger nuclear family sizes reported higher levels of emotional closeness to network members. In the absence of family support, friends would provide moral support (Lang, Staudinger, and Carstensen, 1998). These findings give alternatives to individuals who care about planning and experiencing more successful aging. People in general are aware that in later life, individuals are more likely to experience socio-emotional selectivity. Therefore, it is important that everyone be equipped or work in the direction of securing the social context, such as close friendships and family.

Personality characteristics, on the other hand, have been observed to influence the size of social networks but not the level of emotional closeness with network members. It was found that people who are sociable, energetic, and appreciative of new experiences are more able to maintain a larger social network size than other very old people who are less open to new experiences. Although extroverts may have a larger social network than introverts, this fact does not seem to influence emotional closeness in later life, whereas social context does influence both the size and level of emotional closeness in the network. A more detailed explanation of the influence of personality was mentioned earlier in this book.

The availability of a nuclear family contributes to feelings of social embeddedness but less emotional closeness to others when a spouse or child is available. However, when the nuclear family is not available, the level of emotional closeness to others is even more strongly associated with social embeddedness (Lang, Staudinger, and Carstensen, 1998). Detailed information about the family and children is provided in the next chapter of this book.

Family and Children

Good family support is crucial in older age; a good quality relationship between the child and old parents or grandparents helps them cope with aging. Having a good relationship with their children in later life and hoping or believing in children's support in old age was associated with higher self-esteem and better life satisfaction (Jik Joen Lee, 2005).

> "Home is more than a symbol of quality of life at all ages. It is the place to cover one of our basic needs, namely accommodation. It can have certain benefits for one's physical health and psychological welfare" (Perez et al., 2001: 174).

Family emotional support was significantly associated with depressive symptoms. Individuals who had higher levels of emotional support showed lower symptoms of depression. Older people living with their families were more satisfied because even when they were confined to bed, they could speak with their family, take pleasure in their descendants' safety, and be respected by them. In contrast, the remaining years of the elderly in the USA might be more difficult compared to those in Japan because of their culture of living alone during late adulthood (Shinkawa, Yamaya, Ohrui, Arai, and Sasaki, 2002). Living arrangements in later life are important for coping with aging. In the USA, for instance, it was reported that elderly people tend to live with their spouses only or alone, whereas in some other countries like Japan, elderly people are taken care of by their descendants. Studies showed clear results that elderly people taken care of by family members have higher levels of emotional support, hence better life quality. Consistent with these studies, Laukkanen, Karppi, Heikkinen, and Kauppinen (2001) conducted research to measure the daily activities of older people in relation to living arrangements. The researchers assessed 5,652 participants of both genders aged 65 and over. Participants were selected from two different care settings; participants getting home care and others getting institutional or nursing home care. It was found that people cared for by their home family members were better in activities of daily living than those who were in long-term institutional care. In this study, instrumental activities of daily living (IADL), such as driving, visiting, shopping, and other external needs, were important to maintain the physical health of elders. These needs are more likely to be fulfilled in home care because family members can provide this instrumental support for them. It can be said that since institutions are less likely to provide older people with instrumental support, people decline in their physical activities of daily living. People living in institutions were usually regarded as being lonely, poorer in health, lacking close social networks, and weaker in memory. It was reported that older people living in their homes with family members scored higher in functional abilities than those elders who were in long-term institution care (Laukkanen, Karppi, Heikkinen, and Kauppinen, 2001).

Furthermore, studies found that higher incomes, better health, less loneliness, higher education, and the number of children were important resources for continuous growth and moderating the effects of physical decline and social loss. Individuals who had children in later life reported being more psychologically satisfied than those who had no children.

Having children or not having children in later life was crucial for older people (Blanchard-Fields, Stein, and Watson, 2004). Subsequent studies suggest that even the number of children could influence adjustment in later life. Older individuals who had more children were more satisfied and happier than individuals who had fewer children (Connidis, 1989).

The reaction of family members towards elders was an essential ingredient in making the older people feel respected and satisfied. Being kind, polite, and willing to listen to elders were attitudes that were appreciated by old people.

"An elderly person at home [is like] a living golden treasure." – Chinese saying.

In the interviews conducted with elderly individuals in Kosovo, a recurring grievance expressed by many was the lack of acknowledgment and consultation by their loved ones regarding significant family decisions. They experienced unhappiness and a sense of worthlessness when they were excluded from discussions related to purchasing, selling, making changes, or traveling. This sentiment can be understood by considering the authority and respect that elderly people held during their youth. In the past, older individuals in Kosovo were typically regarded with high esteem and held positions of authority within the family structure. They were often consulted, and their opinions were valued when it came to important matters. Therefore, the shift from being actively involved in decision-making to being disregarded can be profoundly disheartening for the elderly population. The feeling of exclusion from family decisions can have various detrimental effects on the well-being of elderly individuals. It can erode their sense of identity, self-worth, and purpose. Being left out of important discussions can also create a sense of social isolation and marginalization, leading to increased feelings of sadness and diminished overall life satisfaction. Recognizing and addressing the concerns of elderly individuals regarding their involvement in family decisions is essential. By valuing their experiences, wisdom, and contributions, we can promote a sense of dignity and inclusivity for the elderly population. Engaging them in discussions and seeking their input can foster a more positive and fulfilling environment for everyone involved, while also honoring the valuable role they have played in their families and communities throughout their lives.

Respect

The issue of respect for elderly people has been gaining increased attention from gerontologists (Sung, 2004). This trend may reflect a concern over the declining consensus on respectful treatment of elderly persons. Studies have shown that some adults have a propensity to mistreat and abandon elderly people who are sick or weak by neglecting and disregarding their issues. They tend to depict ageism through negative connotations and portrayals of older persons by language, humor, songs, and other forms of art (Sung, 2007). These negative behaviors make elderly people feel worthless to society, which may have serious negative effects on their psychological well-being.

"The perception of being considered worthless by younger generations is a source of deep sadness for elderly individuals."

Without respect, a society cannot treat the aged with decency, have a good attitude toward them, or incorporate them into family and community. According to research, elderly people who are respected tend to have higher life satisfaction and self-esteem, which in turn increases their sense of usefulness and involvement in their families, communities, and close relationships. In their later years, in particular, the respect that elderly individuals receive from others will have a significant psychological impact on their quality of life (Noelker & Harel, 2000).

According to a number of theories, older adults are less socially integrated than younger ones because they are pushed out of society's mainstream by modernization, younger generations, their preference for seclusion from society, or because they are more selective about their social contacts.

In the words of Kofi Annan, Secretary-General of the United Nations:

> "Trees grow strong over the years, rivers wider. Likewise, with age, human beings gain immeasurable depth and breadth of experience and wisdom. That is why older persons should not be only respected and revered: they should be utilized as the rich resource to society that they are." (Powell, 2005).

Before we discuss the value of older people, it would be necessary to repeat some statistics prepared by Simon Powell (2005). The statistics show that the world is getting older because the number of people over the age of 60 will triple over the next 50 years, growing from approximately 600 million to nearly 2 billion. This is due to a combination of factors including a significant decrease of birth rates (below replacement level) and an increase in longevity. Currently, one in every ten persons is 60 years and older; by 2050, this number will grow to 1 in 5. At that point, nearly 80% of the world's older population will be living in less developed regions of the world.

It was suggested that the more compassion and personal vitality young people attributed to elderly people, the older adults would be respected and not avoided. It was also suggested that higher respect for older people and lower avoidance of older people would be associated with higher life satisfaction. Some studies have shown that the way older people are treated by the younger generation and peers as well can have a significant impact on their reports of life satisfaction and well-being (Cheng and Chang, 2006; Powell, 2005). Cheng and Chang (2006) found that respect for elderly individuals had a stronger positive impact on psychological well-being than sociodemographic factors including age, gender, education, and marital status. Paying respect to elderly individuals was a stronger predictor of psychological well-being than independently performing daily activities including shopping for groceries.

Therefore, it is recommended to treat older adults fairly and with dignity regardless of disability or other status, because it was reported that polite talk from young adults could help them maintain and enhance psychological well-being. Respecting old people the way they expect will help them cope with aging. The purpose of this book is to ensure that older adults are treated fairly and with dignity until the end, to respect them and to fulfill their needs when necessary in all stages of their life (Powell, 2005).

391 older people from six European countries were examined. The participants were aged 60 and older and from

different levels of educational, social, and economic backgrounds. These participants were examined to find out their views on human dignity in their lives in association with self-esteem and well-being. When a person was pain-free, washed, in a clean bed and clothes, and had contact with other people, this made him/her feel cared for and well respected. Feeling cared for and respected in later life enhanced the self-esteem, self-worth, and well-being of elders. Understanding the dignity of later life, and applying it to older people by respecting their needs and communicating with them enhanced their self-esteem and well-being. In other words, old people were more satisfied and had more positive perceptions of aging when they felt cared for and respected by others (Bayer, Tadd, and Krajcik, 2005). The way the younger generation or peers responded to the older generation was reported to have a significant impact on life satisfaction and psychological well-being in later life. The more politeness young adults showed to older adults, the more benefits elders gained in life satisfaction and psychological well-being. In this way, older people reported being more respected and not avoided by their peers and younger generation. It was found that politeness and positive communication were associated with life satisfaction and psychological well-being in later life (McCann, Dailey, Giles, and Ota, 2005). Based on the above studies we conclude two significant lessons. The first lesson is for elderly people to gain respect from others by appreciating the younger generation and knowing their own role in society, as suggested in the personality section about the importance of being or enhancing the open and agreeable traits of personality. The second lesson is for the younger generation to pay attention to elderly individuals in terms of their needs and social status. One of the most significant factors influencing elderly individuals was maintaining the respect elderly individuals possessed before. To maintain the respect in post-retirement, elderly individuals must be more open and tolerant with the younger generation. Both the elderly and the young individuals need to be respected for their contributions. Respecting elderly individuals contributes to greater life satisfaction, improves their sense of usefulness and involvement in society. This is important not only for the elderly, but for society in general.

Health

People in later life showed some increase in health-related problems. Among elders aged 62 and over, perceived health declines, preexisting health worsens, and the number of visits to doctors increased as they aged. Having health-related problems in later life contributed to quickening the decline of mental and physical functioning. It was very rare to find a person aged 80 or older to be totally disease-free. Two out of every five people aged 65 to 75 had some kind of physical impairment such as hearing loss, vision impairment, or heart disease (Hayer, Rybash, and Roodin, 1999).

The World Health Organization (WHO) (1948) defines health as a state of complete physical, mental, and social well-being, with an emphasis on physical health and functioning ability. Fakouri and Lyon (2005) define health as a subjective phenomenon manifested as an experience of wellness or illness based on individual evaluations of how they are feeling or doing. Therefore, health-related outcomes encompass both perceived wellness and perceived illness (Fakouri and Lyon, 2005). Fakouri and Lyon (2005) reported a positive relationship between health and life satisfaction. Health was found to be associated with life satisfaction and quality of life (Didinoa, Frolova, Tarana, and Gorodetski, 2015). Poor health was associated with a less active life, low life satisfaction, less happiness, and more loneliness (Jik Joen Lee, 2005).

The most important physical and material resources for the elderly was health; it was found that for a healthy person, it is much easier to cope with the aging process. Further, health was found to be more important in influencing the aging process than the actual age (Fakouri and Lyon, 2005).

Physical health

Physical health in later life is strongly associated with wellness of aging. Rowe and Kahn (1987) have suggested that, to age well, physical health is a very important factor (Strawbridge, Wallhagen, and Cohen, 2002; Withall, Stathi, Davis, Coulson, Thompson, and Fox, 2014). Individuals with good physical health in later life had more positive perceptions of aging than individuals who had decrements in physical functioning. According to Rowe and Kahn (1987), aging well or successfully referred to people who showed little or no functional limitations or disabilities. To age successfully, one had to avoid risk factors like smoking and obesity, and the lack of engagement in productive activities (Strawbridge, et al., 2002). Being connected in social networks also enhanced successful aging. Shmidt (1994) defined successful aging as minimal interruption of usual function. Whereas Baltes and Carstensen (1996) reported that an individual ages well when he does the best with what he/she has, no matter if some minimal signs of disease are present.

Most of the definitions of successful aging, more or less, overlap with Rowe and Kahn's definition. When older people were in good health, they were reported to age well, thus successfully (Strawbridge, Wallhagen, and Cohen, 2002). Traditionally, people relate decline in health to aging. However, many studies report that decline in health was not always a sign of aging. There may be other factors that worsen health and functioning in later life. Lifestyle was found to play an important role in maintaining good health. Elders who regularly exercise and take good care in daily activities were less likely to report physical health problems. Receiving social support predicted fewer health problems and less limited functioning, which were linked to better well-being. However, social conflict was associated with a decline in physical health. It was found that functional limitations in later life were not determined by health or age only. Other factors also play an important role; lifestyle, physical exercise, and social support were also contributors to improving physical health (Seaman and Chen, 2002).

Strawbridge et al. (2002) reported that the study by Rowe and Kahn (1987) had productive consequences for the older population because they traditionally attributed many of the physical problems to age-related deficits. Therefore, with Rowe and Kahn's study of aging, people no longer could consider all age-related deficits as inevitable concomitants of old age. Changes in lifestyles that could improve the well-being of older people were encouraged along with a shift in focus from those doing poorly to those doing well. In other words, people benefited from the study of Rowe and Kahn (Strawbridge et al., 2002).

Turner et al. (2006) conducted research on 3,940 participants over 18 years of age. The main objective of this study was to find out how physical health was associated with mental health across age, gender, and race. This study found that people with physical problems were reported to be more depressed and more psychologically distressed than those individuals who reported no physical problems. On the basis of representative samples across all age and gender categories, participants with physical limitations, in comparison to those who reported no physical limitations, were found to be at a higher risk for depressive and anxiety symptoms. The impact of hearing loss among older people was studied. 2,688 participants aged 53-97 years old were examined. It was found that there was an association between hearing loss and quality of life. Hearing loss had an impact on their activities of daily living. The participants with hearing loss reported communication problems. This communication was a very essential aspect of daily living. It was concluded that hearing loss was significantly associated with low life satisfaction and decline in activities of daily living (Dalton, Cruickshanks, Klein, and Klein, 2003). Low vision was also another cause of low quality of life among older people. Sperazza (2001) conducted research to find out how low vision affected the individual's life. Low vision in this study was defined as having mild visual dysfunction in which their sight was reduced to the extent that their daily activities were reduced. People with low vision avoided social gatherings and isolated themselves because they felt like they were pitied by others and put in different categories of people, in the category of blind people. People with low vision reported having a difficult time coping with aging. They reported having discomfort and lack of confidence to join social gatherings. Many of them reported having bad times and were angry about getting older but not better. It was concluded that low or impaired vision was associated with low self-esteem and low self-worth (Sperazza, 2001). Having a physically active lifestyle can allow older adults to fulfill the need for time structure, meet other individuals with similar passions, and maintain close relationships with families and friends, all being important for life satisfaction (Poon and Fung, 2008).

All of the above studies agree that good health predicts better psychological wellbeing and life satisfaction. However, there are cases when elderly individuals complain about health in late adulthood. Naturally, people with low quality of health complain more than individuals with good health about their aging process. There are other factors that can balance their life satisfaction or psychological well-being when experiencing physical health limitations. But, the real question is, what did elderly people do when they were younger to maintain their good health? How much care was taken about eating and drinking habits, smoking, daily exercises, etc.?

"Everything you do, everything you say, every choice you make when you're young comes back to you later."

If everyone agrees that good health is associated with more successful aging, then everyone must also know what to do during the process of aging. Stop smoking as soon as possible. During my interviews with elderly people in Kosovo, smoking elderly individuals justified their smoking by comparing themselves with other elderly smoking individuals. Just because some elderly people smoke and yet describe themselves as "healthy," does not mean that they would have been healthier in other dimensions of health like teeth, voice, vision, hearing, skin, etc. Therefore, a big majority of researchers confirm that non-smokers have better health than smokers in general. Smoking is just one of the behaviors you can do to maintain your good health. There are other actions you can take to prevent health decline. Eating and drinking habits, for instance, play a crucial role in enhancing and maintaining healthy aging. Unhealthy eating habits are the leading causes of loss of independence as you age. And being independent during late adulthood is a very important factor of life satisfaction and psychological well-being. We mentioned before in this book that daily activities must not discontinue as you get older; you must continue with exercises and activities that are good for health. It is well documented that when elderly people disengage and isolate themselves from society and other activities, they experience a decline in physical health as well, hence a decline in life satisfaction. These behaviors are just a few actions that elderly individuals or even younger adults can take to enhance or maintain their healthy lives. Especially when they all know that good physical health is the leading force to successful aging.

Mental health

Our mental health affects how we think and feel, and how we cope with life's ups and downs. As we move through different stages of life and our circumstances change, our mental health can change too. Despite the increased interest in caring for elders, little treatment is given to them. Mental health refers to the absence of mental disorders and the ability to deal effectively with life events. Accurate perceptions of reality, capacity for independence, successful social relationship and positive self esteem were indicators of a mentally healthy person in later life (Hayer, Rybash and Roodin, 1999).

Prior to understanding the relationship between mental health and successful aging, it is important to understand that older adults who have a positive attitude toward their aging are less likely to be exposed to mental illnesses compared to those with a negative attitude toward their own aging. The majority of studies confirm that attitudes towards aging predict positive mental health. Therefore, it is very important to understand that perception of aging depends on how individuals interpret aging in their own head. If you view aging as a universal process of change and maturity, you will most likely have a more positive experience. On the contrary, if you perceive aging as an outcome rather than a process, you will most likely interpret it as decline in health which leads to a negative experience of aging. Interpretation of events in later life predicts the quality of mental health. This suggests that elderly people should not compare their current health with younger generations but change their attitude towards aging and seek help when needed. With positive perception of aging, elderly individuals are able to avoid symptoms of negative mental health such as depression and low psychological well-being. Positive perception of aging leads to positive mental health, and positive health helps you adapt and continue with daily activities. Njegovan, Man-Son-Hing,

Mitchel and Molnar (2001) interviewed 5,877 participants aged 65 and older to find out if mental health affects functional abilities such as Activities of Daily Living (ADL) and Instrumental Activities of Daily Living (IADL). It was confirmed that mental health was found to be a very important factor to maintain functional abilities such as Activities of Daily Living (ADL) which include dressing, eating and Instrumental Activities of Daily Living (IADL) such as working around the house, shopping, banking and cooking.

It is important to understand that mental health problems like depression, stress or anxiety cause individuals to disengage from leisure activities as much as disengagement and isolation can cause mental health problems such as depression, stress or anxiety (Jaewon Lee and Allen, 2021). One way to reduce depressive symptoms and increase psychological well-being in later life, older adults are strongly encouraged to participate in daily activities and develop a positive perception of aging by interpreting aging as change and not as decline. When your attitude towards aging is positive, your mental health will also be positive. As we explained before, when elderly people protect their mental health while aging, the chances of experiencing successful aging are higher.

Mental health was essential to keep the elders active, for example going shopping, working around the house, paying the bills and many other activities or instrumental activities of daily living. Because, decline in mental health is strongly associated with decreases in ADL, IADL and increase in major symptoms of depression. It is one of the major fears for elders to become functionally disabled and dependent on others.

Another factor that is found to affect the quality of mental health among elderly adults is worrying about the future. Fakouri and Lyon (2005) conducted this research to see how worry affected two important components in the quality of life among older adults; life satisfaction and perceived health. In this study, worry was defined as thinking about the future, thus being mentally disturbed and emotionally disturbed which could influence the person's subjective well-being (Fakouri and Lyon, 2005). Worry contributed to lower life satisfaction, negative emotions, physical discomfort, and decreased functional ability. Given the symptoms associated with aging, older adults are subject to several situations that may trigger worry. Different chronic diseases and worries may influence older individuals' life satisfaction or subjective health (Fakouri and Lyon, 2005). Worry was more prevalent among individuals 65-74 years of age and less prevalent among younger individuals. It was concluded that worry about the future among elderly individuals contributed to lower levels of life satisfaction, increased negative emotions, physical decline and decreased functioning ability (Fakouri and Lyon, 2005).

"Spending energy for the unknown future prevents you from enjoying the present"

How to prevent worry then? emphasize on your current situation and engage yourself in behaviors that lead you to better mental health. Because worrying about the future will not bring any good in your life except stress and depression, which prevents you from enjoying the present. Therefore, people who expect that with aging comes health decline are more likely to be worried about their future. These groups of people that associate aging with health decline or even isolation are more likely to accept health decline as normal and refuse to seek help for treatment. Not seeking help is critically harmful for successful aging. These associations were reported by the majority of elderly adults. Many interviews with elderly people confirmed that older individuals expect a decline in health and functioning ability as they get older. The perception of health in later life influenced their life satisfaction more than the number of diseases they had. Thinking about the future, and getting emotionally disturbed, thus worried, was more prevalent among adults aged 65-74.

Independence

It is very unfortunate to notice in many studies and interviews that over 50 percent of elderly people associate aging with dependence and health problems. In a study with over 500 elderly people, it was found that the majority of them did not expect or believe they could have successful aging as defined by Rowe and Kahn 1978. Successful aging, according to Rowe and Khan, refers to having a low probability of disease, high cognitive and physical functional capacity, and active engagement with life. In this study, most elderly participants expected that with aging comes pain, depression, and lower quality of life. And this is a very serious issue because expecting to have these problems in adult life means that they are less likely to believe they need health care, or engage in social activities. Therefore, loss of independence in later life was the major fear of older people. This book encourages all people, especially elderly individuals, to not associate aging with health problems and disengagement from society and activities. Instead, think of aging as a universal process and that many age-related conditions are preventable with good medical care, improved lifestyle, eating habits, social activities, and most importantly, positive aging perception.

Elders wish to have the ability to satisfy their own needs in later life. They wanted to be independent and have control over their lives for as long as possible. For American elders, the main challenge was to remain independent in later life. American people are living longer and growing older during the last century. It was found that 77 million American babies were born during 1946 to 1964, which would mean they have reached the age of 75 by the year 2018 (Braunstein, 2004), and it is expected that they will live for nearly 10 more years. As people live longer and grow older compared to before, they are more likely to suffer from chronic illnesses such as visual impairment, hearing impairment, and other physical and mental health problems. These health problems would lead to some loss of independence among elderly people. In addition, in another study, it was suggested that older people in America were dependent on public transportation and community support, thus engaging in community activities. If these needed elements were not provided for them, they were more vulnerable to feelings of isolation, depression, and other physical and mental problems (Braunstein, 2004). Individuals who are isolated often develop feelings of hopelessness and depression, and the negative effects on their mental health can lower their quality of life.

Being functionally and cognitively dependent in later life was the most challenging aspect for the elders. The ability to control their own lives by themselves enhanced their self-esteem, well-being, and perception of aging. Many old people fear becoming a burden on others, increasing infirmity, disabling illnesses, the need for help with personal care, and loss of close social networks because these were rated as factors that threatened independence and dignity (Bayer, Tadd, and Krajcik, 2005). To an extent that older women reported preferring death rather than being dependent or a burden on other people in terms of health (Luntz, 2000). The role of the younger generation, close relatives in particular, must understand that elderly people are not a burden, on the contrary, they are needed in the family and society. Make sure your elderly loved ones are able to maintain their sense of self in the environment they live in. Offer them a sense of control, because in some cases, independence is the only thing elderly individuals feel that they can control. Independence gives the elderly a sense of purpose in their life.

Because losing independence in later life was associated with mental disorders (Njegovan et al., 2001). It was reported that 30 to 35 percent of older adults aged 71 and over depended on others for their activities of daily living (Thomas, Williams, Richardson, and Tinetti, 1996). Thomas et al. (1996) examined 1,103 elders aged 71 and older to investigate the relationship between dependence for activities of daily living and physical and mental health. It was reported that dependence was an important predictor of lower physical performance and cognitive impairment in later life. And it was concluded that an increased level of functional dependency was significantly associated with worsening of physical and mental health. Physical health and mental health are often interrelated; they predict one another. If one of them decreases, the other one is expected to decrease as well

(Thomas et al., 1996). Therefore, while we cannot avoid some barriers to independence, we can take the time to understand the importance of independence in elderly individuals and look for ways to increase opportunities for independent living. For instance, encourage and let elderly people decide for their own actions. Because removing the sense of self-determination can lead them to feelings of depression.

It is not right to assume that elderly people are unable to make a decision for themselves. On the contrary, give them enough time to make decisions, encourage them to make healthy life choices, and ask their opinion on major decisions, especially those that concern them. In addition, we must keep in mind that in most cases, to be independent during later life depends on your lifestyle during young adulthood.

Religiosity

In many cultures, religious behavior is believed to have the power to moderate emotions. Therefore, many people consider religion as a source of coping in later life. Dein and Stygail (1997) reviewed a number of studies that examined the use of religion in coping with chronic illness. In this study, it was reported that religion helps older people deal with aging and that it was positively associated with better adjustment after retirement.

There has been a recent suggestion that religious beliefs or involvement may compensate for a lack of close social relations for some people. Religious beliefs and participation in religious activities are closely related to positive aging outcomes, particularly life satisfaction and the absence of mental disorders. Many problems confront older persons that can be linked to their sense of spiritual well-being. Religion helps in dealing with difficult times, for example, coping with the death of a spouse, relative, or friend. The death of a spouse or close friends for religious people was perceived as God's law (Strydom, 2005). Religion can change the perception of older people in dealing with stress. It can shape the cognitive appraisal and the appraisal of personal resources available to respond to stress (Pargament, 1997).

Spirituality and religion have been found to play an important part in many older people's lives, and they have been found to be strongly associated with physical health, mental health, well-being, and lower mortality levels.

Many other studies have also found that spiritual beliefs influence psychological well-being at older age. Kirby et al. (2004) examined psychological well-being when health declines and people become weak. They examined 233 older adults, 60 males and 173 females, aged 65 and over. Participants included in this study could hear, understand, and were able to respond to the questions. It was reported that weak health had a negative influence on psychological well-being. On the other hand, spirituality and religion were found to be very helpful for maintaining psychological well-being (PWB), especially for individuals with higher levels of frailty. In this study, it was concluded that spirituality was a moderator to reduce the negative effects of frailty on psychological well-being. In other words, spirituality helps the elderly maintain their psychological well-being despite the health issues they might have.

Almost all the studies regarding spirituality and religion are consistent in their conclusions that spirituality and religion benefit the well-being of older adults (Kirby, Coleman, and

Daley, 2004). As people age and their health deteriorates, they are in a position to confront their own mortality and have to cope with the increasing challenges facing them, which seems to have a negative impact on well-being unless they have other resources that they could rely on. In this case, spirituality and religion were found to play a very important role in challenging the effects of aging in later life. Even though in some other studies it was suggested that as older people age and approach death, their religious activities decrease, their religious feelings are stable and even increase as they age. Religion and spirituality were reported to be associated with better physical health, mental health, and well-being. Individuals with higher levels of religiosity reported being healthier and with a better quality of life (Krause, 2002).

A group of researchers tried to explain the relationship between religious involvement and psychological well-being and life satisfaction. They also hypothesized that religious involvement is strongly associated with mental health. In this study, it was reported that religious involvement protected people from stressful events. Alcohol, gambling, marital disharmony, divorce, and deviant society are all discouraged by most religions, thus individuals who practiced their religion would avoid these harmful behaviors. One of the reasons was reported to be the fear of divine punishment or hellfire (Edison, Boardman, Williams, and Jackson, 2001). In certain religious activities, songs, shouting, and physical activities are involved as part of their worship style. Individuals who practiced these kinds of activities reported being healthier mentally and were more emotionally satisfied than those who did not practice. Those who were regular members of religious congregations reported having larger social networks and having more interactions with other people than those who were not regular. Having larger social networks and interactions with people helped old individuals cope with aging (Edison, Boardman, Williams, and Jackson, 2001).

Religious involvement increases feelings of self-esteem and self-worth. Prayer, for instance, was reported to reduce distress and at the same time enhance feelings of well-being because through prayer people reported being able to maintain the relationship between themselves and their perceived 'divine other'. Through these practices, individuals gained spiritual satisfaction and by repeating them, they strengthened their spiritual experiences, feelings of hopefulness, optimism, reduced negative emotions, and found peace in their hearts. Religious involvement helped reduce the effect of stress or any other personal difficulty. For example, in America, people who were facing problems or difficulties turned to their religious prayers or other religious activities, hoping for help to cope with problems. Religious involvement was consistently found to have a strong association with psychological well-being and life satisfaction (Edison, Boardman, Williams, and Jackson, 2001).

Age

Let us start this section with a short story:

"An 80-year-old person goes to the doctor and complains about his knee; after the examination, the doctor tells him that

the knee problem is caused by old age. 'You are now 80 years old,' said the doctor.

The old man responded, 'My other knee is also 80 years old, but it does not hurt!'"

The wisdom behind the story is that some health issues are not necessarily caused by age itself, but other factors can cause health issues as well. For example, the way people sit or exercise may harm one knee more than the other one.

Population aging is one of the main demographic trends in developed countries. Thus, in 2017, the share of the population aged over 60 years was 25% in Europe, in Russia -21%. According to UN forecasts, by 2050, the share will grow and reach 34% in Europe and 28% in Russia (United Nations 2017).

Many people believe that old people are unhappier than young people. However, as we move on with literature and many studies, we come to understand that it is not the age itself that leads to unhappiness. On the contrary, many studies showed that when controlling other sociodemographic factors like health, family, and income, older individuals were reported to be happier than younger individuals (Kutubaeva, 2019). In general, actual age was found to be a very important factor influencing an older person's ability to cope with aging. Probably because of the short amount of time left for them to live, this is what most people would think of. However, since the time left for them is not certain, then other factors come in, like personal resources to influence their coping with aging.

As people age, they come to realize that life is finite. Grown children leave, grandchildren are born, losses like widowhood or death of a partner or close friend more or less indicate the ending of life. When this time comes, older adults are no longer interested in long-range goals like education or ambition but are more interested in short-range goals or immediate benefits such as emotional gratification. In other words, as people age and time is running out, their achievement goals shift from long-range to immediate benefits because they come to realize that they are not going to live forever. This indicates that as people get older, life satisfaction or happiness should not naturally decrease but change the format and take actions to maintain the level of life satisfaction and happiness. How to do this? People must understand that self-assessment of health and functional capacity in daily activities are the strongest explanatory factors of life satisfaction. It is the perception or the interpretation of one's physical and mental health that affects the quality of life satisfaction in later adulthood (Berg, Hassing, McClearn, and Johansson, 2006). It is expected for health to weaken as one gets older, but that should not weaken one's life satisfaction or happiness. Life satisfaction or happiness after retirement does not decrease because one is getting older; it might decrease because of other factors that one took for granted such as the social network, health, family, etc.

Krause (2005) investigated 1,315 older people. Preliminary analyses revealed that the average age of these study participants was 78.6 years; 39% were men, 50% were married at the time the interview took place. In this study, Krause reported that people with physical and mental limitations showed higher levels of stress, and he concluded that age-related deterioration of physical and mental health may make older people more vulnerable to stress. On the other hand, it could be that as people grow older, they encounter more physical and mental problems, which make their life more difficult and stressful (Krause, 2005). These studies indicate that it is not the age that makes elderly people more stressful but the deterioration of health. Therefore, we can say that if elderly people take care of their health, their level of life satisfaction will not drop. On the contrary, many studies in Europe found that elderly people are happier than middle-aged individuals (Gwozdz and Sousa-Poza 2010; Frey and Stutzer, 2002). Life satisfaction among middle-aged individuals was weaker than that of those above the age of 60. There are a few possible explanations for the reason why some elderly people might experience less stress than middle-aged individuals. Frey and Stutzer (2002) mentioned that one of the possible reasons for the elderly to have a positive relationship between aging and happiness is the level of expectations and aspirations. As one gets older, they realize that compared to younger generations, older generations are more likely to retire from work, and they cope easier with the loss of loved ones. It is about the perception and interpretation of events that are more likely to happen in later adulthood like the loss of loved ones and jobs. It is the decision of the person how to interpret events, in a positive or negative way.

Another reason for some elderly people to continue to have happy and satisfied lives is to revise their targets or goals. Set up more achievable targets as compared to when he or she was younger. This factor is related to earlier life achievements; people who consider themselves as successful might cope better than unsuccessful people. Therefore, it is very important how you perceive your past life experience.

As we mentioned in the beginning of the book, for more successful aging, individuals must prepare while they can, while younger. Because it is clearly found that aging itself does not worsen life satisfaction or happiness, negative perceptions about aging affect life satisfaction.

Education and Income

Age was obviously a key factor influencing the view of aging, but research has shown the significant influence of other factors as well. Some of these other factors are health, education, being married, income, and the number of children (Connidis, 1989).

An investigation was conducted with 4,034 Germans aged 40-85 to find out about the personal experience of aging and the consequences for subjective well-being. In other words, the study intended to find out how their perception of the personal aging process was related to indicators of subjective well-being (Blanchard-Fields, Stein, and Watson, 2004). Three dimensions of aging were used in this study:

a) physical decline, such as the loss of vitality and weakness in health,

b) continuous growth or progression of life, and

c) social loss, such as no longer being needed by others and also decreasing respect from others.

People with higher income and higher education experienced the process of aging less in terms of physical decline and social loss, and more in terms of continuous growth (Blanchard-Fields, Stein, and Watson, 2004), indicating that higher levels of income and education did contribute to successful aging. It is much easier to cope with the aging process when a person is educated and without financial worries. On the contrary, older adults with lower education and income were weaker in health-related quality of life (HRQOL) (Keyes, Michalec, Kobau, and Zahran, 2005).

Although education and income similarly contributed to better quality of aging, education and income contribute to aging quality in different directions. Longitudinal research studies have reported that education plays a greater role in terms of functional limitations, whereas income has a much stronger effect on their progression or course of life (House, Lantz, and Herd, 2005). A higher educational level, for instance, leads to better ability to deal with aging because it was found that people with higher education have less social loss and more opportunity for continuous growth (Steverink, Westerhof, Bode, and Dittmann-Kohli, 2001). Education was strongly associated with onset or maintaining health in good condition by avoiding harmful products, whereas income was associated with the progression of the life course (Zimmer and House, 2003). Individuals with higher education were reported to avoid anything harmful to subjective well-being, such as smoking and lack of exercise, lack of social relationships, psychological orientations such as anger/ hostility, or lack of efficacy/ control. Although education is an early experience, it has implications for later life because education is considerably associated with the prevention of functional health disorders.

Income, on the other hand, is also important in the prevention of functional health problems, but income is even more important in terms of facilitating improvements and recovery from problems because with higher income, people can purchase better quality health care (Zimmer and House, 2003). Income does not only maintain health but also determines the course of that problem by improving or getting worse. Individuals with higher income could maintain their health in good condition, whereas individuals with low income could not maintain their health. Elders with higher income were able to live in a better environment and have a less stressful living situation. In addition, better income contributed to better psychological well-being in later life.

In regards to income as a factor predicting psychological well-being and life satisfaction in late adulthood, studies have reported that the quality of marriage and relationships with family and close friends will well-maintain happiness even if elderly individuals have no high income. Although income is important in making retirement plans, relatively small changes in income may have little effect on retirement satisfaction. Job satisfaction among older workers is not predicted by income (McNeely, 1988). This is confirmed with studies indicating a very strong relationship between good quality of marriage and happiness in later life, indicating that everyone must work on having good relationships with family and close friends for better life satisfaction. When speaking about good family relationships, researchers have suggested that these relationships need to be strengthened during young adulthood and maintained or improved during older adulthood.

The National Health Interview Survey (NHIS) suggested that American people who possessed higher levels of education and higher income were more likely to experience compression of morbidity and functional limitations. Compression of morbidity is a term that means reducing the length of time a person spends sick or disabled. This study claims that higher education and income together lessen the chances of being unhealthy. However, those who possessed lower or no education and lower income manifested declines in health and functional status over their entire adult life course (House et al., 1994).

Coppin, Ferrucci, Lauretani, and Phillips (2006) studied the consequences of low socioeconomic status, including education and income, and went even deeper to determine which physiological aspects are affected by low education and income. Their sample consisted of 1,025 individuals over 65 with a mean age of 75.5 and SD = 7.3. Coppin et al. (2006) measured physical performance by asking the participants to complete the Short Physical Performance Battery (SPPB) by walking 20 laps of 20-meter distance as quickly as they could. The Short Physical Performance Battery (SPPB) is an objective assessment tool developed by the National Institute on Aging for evaluating lower extremity functioning in older persons. Results showed that people with lower education and income scored lower on the Short Physical Performance Battery (SPPB). This result was consistent with the past studies that have reported a strong association between factors like education and income, and physiological functioning (Coppin, Ferrucci, Lauretani, and Phillips, 2006).

Cagney and Lauderdale (2002) examined the extent to which education, wealth, and income influenced psychological well-being and cognitive functioning. They interviewed 6,577 participants aged 70 and over. This study reported that higher levels of education, wealth, and income were significantly associated with greater cognitive functioning and psychological well-being in later life.

Education and income were not found to have a similar effect on cognitive functioning. Cognitive functioning refers to the performance of the mental processes of perception, learning, memory, understanding, awareness, reasoning, judgment, intuition, and language. The correlation between education and cognitive functioning is more significant than the relationship between income and cognitive functioning. This suggests that it is crucial to prioritize obtaining the required education as early as possible. Education is not solely connected to formal qualifications or accolades; instead, it encompasses the facilitation of learning and the acquisition of various elements, including knowledge, skills, values, morals, beliefs, habits, and personal growth, throughout one's life, spanning from childhood to adulthood.

Marital Status

Marital status is an important factor determining adjustment to retirement. The spouse is a very vital resource in coping with aging. It was reported that losing a spouse in later life affected the self-esteem of elderly individuals. Losing a spouse in later life lowered the level of emotional support, and as a result, spousal loss decreased the self-esteem of elderly people. The literature suggested that compared to younger individuals, elderly people are more likely to experience physical health problems, social isolations, and a decrease in social activities. Therefore, after all those losses, losing a spouse decreases the self-esteem of elderly people. Loss of a meaningful spousal role and loss of significant relationships with close friends may decrease self-esteem. It was concluded that increased physical problems, social isolation, and decrease in social activities following spousal death may negatively influence perceptions of self-worth and self-esteem (Arens, 1982; Ferraro, 1984). The loss of a spouse is considered an adverse life event, which can greatly impact life satisfaction and requires adaptation by a surviving spouse. Widows, when compared to non-widowed older adults, tend to have a greater decline in physical health, mental health, and social functioning. The transition to widowhood might take some time, and many older adults do not return to their pre-loss levels of life satisfaction. However, some other studies have shown that some widowed older adults improved their quality of life after spousal loss (Bonanno et al. 2002). These contradictory studies indicate that participating in social activities can provide a positive realm for adjustment to negative life circumstances (e.g., the loss of a spouse) by restoring a sense of well-being and social connectedness.

Bandura (1977, 1993) proposed that self-efficacy beliefs of individuals regarding perceptions of health-related quality of early life can contribute to their physical and mental health-related quality of later life. According to this theory of self-efficacy beliefs, it can be concluded that individuals with lower levels of perceived self-efficacy in various domains such as interpersonal, instrumental, emotional, social support, financial, physical health, nutritional, and spiritual health reported lowered perceptions of health-related quality of life, life satisfaction, and self-esteem for both widows and widowers. Based on these results, which showed significant associations between perceived self-efficacy ratings and the ratings of health-related quality of life, life satisfactions, and self-esteem, it can be concluded that the perceived self-efficacy of elderly spouses was a reliable predictor of their later adjustments to spousal loss and widowhood. In other words, the effects of widowhood depended on their pre-widowhood self-efficacy beliefs about life satisfaction, quality of life, and self-esteem (Fry, 2001). As stated above, with some level of adaptation, the majority of elderly widows and widowers will eventually return to their initial levels of life satisfaction depending on how strong their previous life satisfaction and quality of relationships with spouse and others (Lucas, Clark, Georgellis, and Diener, 2003).

Being widowed or a widower is very difficult in older age. Often after the death of a partner, the spouse is more vulnerable to illnesses. Both widows and widowers share some stresses, and they also rate themselves low in life satisfaction compared to married couples. The first problem of a widow or a widower is to cope with stress after losing their spouse, which may take years of grieving. It can be concluded that being married or having a spouse very late in life is very important for a good and happy life (Rogers, 1986). However, an important point was reported in relation to marital status and life satisfaction. It was found that in addition to marital status, the quality of marital status had a stronger influence on life satisfaction among elderly individuals (George, 2010). Therefore, considering that marital relationships cannot suddenly improve as you age, you need to invest and work on building positive marital relationships long before you retire or feel you belong to the elderly group. One of the explanations for this protective effect is social support. Numerous studies have indicated that one's spouse is often named as the most important source of support in the elderly population, especially for health-related issues. Studies suggest that widowhood tends to put individuals at a greater risk for declining health, depression, and other negative outcomes. These negative outcomes of widowhood may lead to decreased social participation or increased self-isolation. As we discussed before, it is the social isolation that negatively influences our life satisfaction and psychological well-being. Therefore, based on previous experiments, when individuals experience the loss of a spouse or a family member, we must either be equipped with a wider social network for social and emotional support. Family members and close friends are shown to help best when dealing with spouse loss. Married or widowed, individuals must avoid self-isolation as they get older; instead, they must engage in social activities. Family members and close friends will be there for you whenever necessary. Also, ongoing involvement in leisure activities is associated with improved well-being and higher levels of life satisfaction in older adults. These suggestions come from previous studies claiming that factors such as social support, leisure participation, and engagement in physical activities can help to sustain older adults' psychological well-being in situations like the loss of a spouse or other close ones (Tomas, Sancho, Gutierrez, & Galiana, 2014).

In conclusion, this chapter of the book emphasizes three main factors for improving life satisfaction and psychological well-being during later life: being married tends to result in better life satisfaction, the quality of marriage rather than status has a stronger influence on life satisfaction and psychological well-being. Lastly, no matter if you are married or widowed, you should never disengage from society and social activities, especially activities you used to do before.

Gender

Gender has been identified as a significant factor influencing adjustment in later life, with women exhibiting certain distinct patterns. According to Hatch (2005), women tend to place greater emphasis on their appearance compared to men, leading to a higher likelihood of holding negative perceptions of old age. Additionally, gender differences in coping with aging have been observed, particularly in relation to depression among older individuals. Research cited by Hatch (2005) indicates that women below the age of 60 experience higher levels of depressive symptoms compared to men. However, after the age of 60, men tend to report an increase in depressive symptoms while women's levels remain relatively stable. These findings suggest that gender plays a role in the manifestation of depression in later life, but the differences tend to diminish as individuals age.

Moreover, as individuals grow older, women demonstrate a tendency to rely more on their close social network for help when facing stressful life events. In contrast, men may seek less support from their social network. Additionally, Krause (1991) highlights that women tend to have larger social networks in later life compared to men. This implies that women have more extensive social connections that they can turn to for assistance and emotional support during challenging times.

Additionally, while men tended to have a larger number of friends overall, women had a higher number of close friends. The presence of close friendships among women offers them greater emotional support, which proves crucial in dealing with stressful situations during later life. In contrast, men included fewer close friends within their social support network. As a result, they received less emotional support and experienced a lower level of respect from their friends.

Moreover, older men received less emotional and health support from their children compared to women. Conversely, women tended to receive more emotional and health support from their children. These findings were reported by Depner and Ingersoll-Daydon (1988) and suggest that men may have a more limited support network in terms of emotional assistance and may be less reliant on their children for emotional and health-related support compared to women.

Further, mothers reported having closer relationships with their children than fathers. The relationships of mothers and children were warmer because women were more likely to be lonely in later life than men. In other words, the chances for widows to get married in later life are lower compared to widowers. Therefore, women develop warmer relationships with children because children are all they have. Some of the possible reasons could be that income for women was lower because they usually worked in lower-paying jobs compared to men. And men, on the other hand, were more likely to remarry when their spouse died, which contributed to less loneliness in later life for men (Long and Martin, 2000). However, when male older individuals faced stressful life events, they used their own planful problem-solving strategies to cope with stress, whereas women were more likely to seek help or support from their social network.

Another gender difference reported was gender-health inequalities. Women's health in later life was found to be positively associated with social relations, whereas among men, this association was not significant. In this study, it was also reported that women had better abilities to cope with stress in later life than men (Prus and Gee, 2003). Krause (2003) further reported that men with cognitive impairment were more likely to seek help from general medical services. They were reluctant to seek help from psychological centers even though they needed psychological help. Such reluctance, though, was not found among women.

In conclusion, the above studies suggest that during the processes of aging, there could be slowdowns and challenges. But, studies have also suggested resources that can be utilized in dealing with life transitions, such as retirement, illness, or widowhood, or with other problems they encounter. Elderly people have the capacity to regain lost functioning and make a successful adaptation to the challenges facing them.

Theories: Aging And Adjustment to Retirement

This section highlights the main theories that explain the psychosocial factors influencing aging and adjustment to retirement. Regarding life after retirement, Havighurst (1986) reported:

> "The decrease of interaction proceeds against the desires of most aging men and women. The older person who ages optimally is the person who stays active and manages to resist the shrinkage of their social world. They maintain the activities of middle age as long as possible and then find substitutes for work when forced to retire and substitutes for friends and loved ones whom they lose through death" (Quadagon, 1999: 161).

Since the 1960s, interest in studying the post-retirement population in terms of their aging image has gradually increased (Featherstone and Wernick, 1995). The number of professional and scientific journals about gerontology has also increased during the last decade. Some possible reasons for this increased number of professional journals would be the increasing number of old people, their increased lifespan, and the ever-growing burden on society for their care (Rogers, 1986). According to Sarah and Pat (1989), by the late 1960s, it was accepted that the normal period of full-time employment would stop for most of the population at the age of 60 or 65. At these ages, most people, especially men, stop working, and as a result, this creates a major crisis of identity. As people age, their occupational and physical status changes to the extent that they come to understand that they are no longer the prime movers within their family or at work. Parsons (1942) described the problems that retirement brings to people by postulating:

> "In view of the very great significance of occupational status and its psychological correlates, retirement leaves the older man in a peculiarly functionless situation, cut off from participation in the most important interests and activities of society…. Retirement not only cuts the ties of the job itself but also greatly loosens those to the community of residence. It may be surmised that this structural isolation from kinship, occupational, and community ties is the fundamental basis of the recent political agitation for help to the old. It is suggested that it is far less the financial hardship of the position of elderly people than their "social isolation" which makes old age a "problem" (Parsons, 1942: 616).

Levinson (1980) suggested that after retirement, people should learn how to pass their leadership to the younger generation because they cannot be the prime movers in the family or at work anymore. As Levinson (1987) reported, old people should take the "back seat" in their later stage of life and let others lead in the family or work. This process was termed as the view *from the bridge.* Levinson suggested that elderly people should pass their duties and responsibilities to the younger generation and learn to agree with the younger generation. It is not necessary for them to avoid all of those cares and duties that they had before, but they could hand over some of those duties and cares to other family members or younger friends (Stuart-Hamilton, 1994). Levinson identified four overlapping stages in the life cycle of people, with each stage lasting about 25 years and each stage having its own psychological, physical, and social issues (Stuart-Hamilton, 1994; and Rogers, 1986). It is the last stage of life, which starts at the age of 60 and over, when individuals experience more life distractions. During this time, older adults experience higher levels of stress and anxiety because of their physical decline and also because they are labeled as *old people* after their sixties. In this stage, people are recommended to accept the realities of their past, present, and future in order to be satisfied with their lives. However, if they come to reject these realities, they would be more vulnerable to psychosocial and physical problems (Hayer, Rybash, and Roodin, 1999).

Old age is usually referred to as the last stage of human life. Erik Erikson proposed eight stages of the life span. Each life stage has specific issues that are generalized to all people who go through the stages. The first six stages include the years from birth to young adulthood. The seventh stage includes individuals aged 26 to 50. This is the stage of *generativity versus stagnation,* where generativity results when individuals try to satisfy their needs by turning towards others, such as bearing children and trying to contribute to the younger generation by mentoring and guiding them. However, stagnation results when individuals at this stage do not get satisfaction from others; in other words, their contribution to others results in frustration or lack of fulfillment (Quadagon, 1999). Regarding generativity, Levinson (1996) reported:

> "Most of us during our forties and fifties become "senior members" in our own particular world. We are responsible not only for our own work but also for the development of the current generation of young adults.... It is possible in this era to become more maturely creative, more responsible for self and others, more universal in outlook, more capable of intimacy than ever before" (Quadagon, 1996: 20).

The eighth and final stage of life starts from the age of sixty and above. This stage of life encounters the issues of ego integrity versus despair. In later life, Erikson argued that individuals who reach the last stage of their lives are characterized by a psychological conflict between ego integrity and despair. Individuals with ego integrity would appraise their old age and will acquire wisdom, acknowledge the universal nature of humanity, and accept the fact of mortality. Here is how Erikson described an individual's ego integrity:

> "It is a post-narcissistic love of human ego---not of the self---as an experience which conveys some

world order and spiritual sense.... It is the acceptance of one's one and only life cycle as something that had to be and that, by necessity, permitted no substitutions.... It is a comradeship with the ordering ways of distant times and a different pursuit.... [but] the possessor of integrity is ready to defend the dignity of his own lifestyle against all physical and economic threats.... Ego integrity, therefore, implies an emotional integration which permits participation by followership as well as acceptance of the responsibilities of leadership" (Belsky, 1990: 268-269).

According to Erikson's theory, the person who has reached ego integrity should be able to accept the past and death (Belsky, 1990). In this stage, people become more aware of the fact that death is approaching. This is the stage when old people look back at what they have done and achieved during their life period. Individuals who perceive that their seven former stages, as proposed by Erik Erikson, were successfully accomplished, were reported to have had a meaningful and satisfying life, which is *ego integrity*. However, some other people reported being not satisfied with their life and not having had a meaningful life during the seven former stages. These people were categorized under despair. These people were not happy with their lives, so they blamed themselves for their disappointment, thus despair (Rogers, 1986). Despairing old people may become withdrawn and hesitant or destructive in their old age. Regarding the stage of ego integrity versus despair, Erik Erikson (1968) reported:

> "A meaningful old age, then.... serves the need for that integrated heritage which gives an indispensable perspective on the life cycle. Strength here takes the form of that detached yet active concern with life bounded with death, which we call wisdom....
>
> To whatever abyss ultimate concerns may lead individual men, man as a psychosocial creature will face, toward the end of his life, a new edition of the new identity crisis which we may state in words, "I am what survives me" (Rogers, 1986: 140--144).

Based on role theory, the main task for men is their work, and with retirement, their main task is finished, so it is this time when disengagement begins (Fennell, Phillipson, and Evers, 1988). As individuals age, members of their social network start dropping off one by one. Their neighborhood also starts disappearing, and as a result, older individuals may give up their social or external resources and instead turn themselves towards their inner resources. Older individuals, in fact, automatically distance themselves from society and external resources. Furthermore, disengagement theorists reported that this disengagement process is universal, normal, and natural (Belsky, 1990). Disengagement has been defined as an inevitable process in which many of the relationships between a person and other members of society are severed, and those remaining are altered in quality (Belsky, 1990). According to disengagement theory, retirement is the time for elders to relax and maintain higher morale in old age; in other words, disengagement suggests that old people should stop getting involved in their daily activities. Disengagement involves the process of social and psychological withdrawal of an individual from society. Disengagement occurs when "the individual becomes sharply aware of the shortness of life and the scarcity of time remaining to him, and if he perceives his life space as decreasing" (Quadagon, 1999: 215). Disengagement usually leads to a reduction in daily activities and ego energy in later life. This view has been challenged by gerontologists who disagreed with disengagement theorists because disengagement theory challenged the traditional view about the importance of maintaining high levels of activity and interaction in old age. The alternate theory that was proposed was the *activity theory,* which proposes that happiness in later life is to stay as involved in life activities as possible (Belsky, 1990). The contradiction between disengagement and activity theory is that, according to disengagement theory, older adults should withdraw from their daily activities and focus on maintaining higher morale and avoid life activities, whereas **gerontologists insist that older people should remain active for as long as possible.**

In one of the largest studies conducted, the Duke Project involved 250 individuals aged 60 and older. It was concluded that older individuals who reported the highest activity levels scored highest in life satisfaction, contrary to what disengagement theory would predict. Disengagement theory would suggest that in older age people who score high in daily activities would score low in life satisfaction (Quadagon, 1999). Gerontologists suggest that, in order to avoid certain psychosocial and health declines, older adults should continue pursuing their usual life activities and not disengage from them (Fennell, Phillipson, and Evers, 1988). Gerontologists suggested that disengagement theory should be rejected because this theory, in some way, forces people to withdraw from society and life activities in older age, which according to activity theory harms the psychosocial and physical health of individuals in later life (Belsky, 1990).

In conclusion, old age was reported to be a very complicated stage whereby older individuals face higher levels of stress, anxiety, depression, and health problems compared to the younger generation. So, they come to face different problems in trying to adjust to those psychosocial and physical problems. Erikson and Levinson proposed theories explaining differences at different stages of the lifespan. Erikson proposed eight stages of life, but we focused more on the last stage which is ego integrity versus despair. Ego integrity results when an old person accepts the past and death, whereas despair on the other hand results when an old person is not satisfied with his past and does not want to accept that death is approaching. Different theorists suggested different ideas about adjustment in later life, such as disengagement theory suggesting that retired people should decrease their life activities and relax. On the other hand, activity theory suggests just the opposite of disengagement theory, which is not to withdraw from life activities but instead keep life as active as possible. Thus, there is both agreement and disagreement on the most appropriate approach to coping with adjustment to retirement.

Social Comparison

Social comparison theory, introduced by Leon Festinger in 1954, suggests that elderly individuals tend to value their own personal and social situation by assessing how they compare to those of similar others. The theory claims that individuals have an innate drive to evaluate their personal, social, and economic situation in contrast to those of other people (Festinger, 1954). These comparisons are made in two directions, namely upward social comparisons and downward social comparisons.

The first direction (upward social comparisons) happens when people compare themselves with individuals or groups who are perceived to be more advantaged than them, and this situation is reported to result in less life satisfaction and more negative outcomes. This kind of comparison (upwards) can be a helpful tool when used correctly, if they use it as a motivational tool to improve themselves. However, in the case of elderly people, many studies have found that it leads to life dissatisfaction.

The second direction (downward social comparisons), on the other hand, occurs when people compare themselves with individuals or groups who are perceived to be less advantaged than them, and this situation is assumed to result in better life satisfaction and more positive outcomes, including increased psychological wellbeing (Beaumont and Kenealy, 2004; Gana, Alaphilippe, and Bailly, 2004). In addition, when elderly individuals perceive themselves as having a better situation than their peers, they are more likely to have a better self-image and take better care of themselves (Kesici and Erdogan, 2010). Downward comparisons will make people feel more satisfied about their own personal and social condition (Frieswijk, Buunk, Steverink, and Slaet, 2004).

Substantial evidence showed that compared to young and middle-aged people, older individuals were more likely to use downward social comparisons. It was found that the use of downward social comparisons explained almost all of the higher social well-being reported by older adults (Beaumont and Kenealy, 2004; Gana, Alaphilippe, and Bailly, 2004). By comparing themselves with others who are in weaker circumstances, people may create a lower reference point to evaluate their situation, which can make them redefine their situation in a more positive way (Frieswijk, Buunk, Steverink, and Slaet, 2004). These kinds of downward comparisons will make people feel more satisfied about their own circumstances and have been shown to be more predictive of life satisfaction than those kinds of upward comparisons. Because, by comparing themselves with others who are in a better situation, people engage in an unfavorable match that will make them feel more dissatisfied with their lives (Diener & Fujita, 1997).

The above studies and results make us understand that when facing difficult situations during later adulthood, it is better that you compare your situation with people having even more difficult situations than yours. By making a downward comparison with others, you will perceive your well-being better and more appreciative than making an upward comparison. There may be cases when an upward comparison could help you boost your life activities, but when facing physical declines during later adulthood, studies found that downward comparison makes you perceive your situation as more advantageous and positive.

Social Integration

The concept "social integration" is often used interchangeably with social support, social networks, social contacts, and social inclusion as opposed to social segregation and isolation. One definition of social integration refers to the degree to which an individual is involved in social participation with others, such as family, friends, relatives, and other communities. One of the most important considerations that may help older adults maintain and improve their well-being and satisfy their psychosocial needs is social integration. Studies have reported that older adults who are integrated into society (social integration) have a better quality of life and are healthier than those who are isolated as they age (Vitman, Iecovich, and Alfasi, 2014).

At the contextual level, social integration refers to the extent to which residential environments are characterized by bonds among community residents, collective efforts to protect or improve the community, and sustained patterns of community-based interactions. On the contrary, studies have found that elderly people who isolate themselves as they advance in age are more likely to complain about health and life satisfaction. Although elderly people may perceive some declines in health during later adulthood, not all of them perceive it as a negative outcome. Some elderly individuals perceive it as a negative outcome and do nothing about it, while others take actions and continue with social integration at a different pace, maintaining their life satisfaction and well-being.

It is well known that some elderly individuals isolate themselves because of health issues, finding health decline as justification for them to stay at home and withdraw from society in general. However, what they might not realize is that self-isolation worsens their health problems by increasing the risk of cognitive decline, chronic lung disease, arthritis, depression, re-hospitalization, institutionalization, coronary heart disease, and stroke (Journal of American Heart Association, 2022). Withdrawal or isolation from society increases the risk of high blood pressure, obesity, heart disease, a weakened immune system, depression, anxiety, and many other health-related problems.

Studies have shown that leading a socially active life and prioritizing social goals in late life are associated with more positive well-being and life satisfaction (Gerstorf, Hoppmann, Löckenhoff, Infurna, Schupp, Wagner, and Ram, 2016). There are many ways elderly individuals can integrate themselves with others. Older adults are encouraged to engage in assisting their children, grandchildren, taking on care responsibilities, performing household tasks, or working as volunteers in the community. Their contributions in providing wisdom and advice to younger generations and society as a whole should be acknowledged.

Older persons should be empowered to pursue their interests and hobbies, building upon life achievements. Interventions may include offering a wide range of activities that are group-based and tailored to specific age groups. Older persons' engagement with friends and relatives should be facilitated where possible. Practical barriers such as communication difficulties, cost, and transport should be addressed.

This book chapter emphasizes the need to consider more actively how to integrate elderly individuals and ensure their participation in social activities.

Psychosocial Resources

In the last two decades, there has been growing interest in how personal resources and psychosocial resources enhance adaptation and quality of life and contribute to the psychological well-being of individuals after retirement. Psychosocial resources refer to personal skills, beliefs, and personality factors that determine how individuals cope with their stressful events. Studies have reported that if individuals fail to improve or maintain the above resources, they are more likely to decline in life satisfaction and psychological well-being. Psychosocial resources could include a vast number of factors depending on the objectives of researchers, but most researchers agree that some of the main factors include optimism, self-esteem, personal control, and social support (Taylor, 2011).

Several psychosocial factors have been hypothesized to directly affect happiness in late adulthood or after retirement. Psychosocial resources, including optimism, self-esteem, personal control, and social support, influence the relationship between the socioeconomic status of people and their health. Some researchers found that the positive perception of elderly people about having full control of their lifestyle has a direct positive impact on positive well-being. In other words, being independent and having control over the outcomes in later life predicts higher life satisfaction and psychological well-being. **Optimism** refers to outcome expectancies that positive things will happen to the self. Optimistic individuals are more likely to experience greater positive outcomes, including physical and psychological well-being, compared to their peers holding pessimistic approaches (Kubzansky, Wright, Cohen,

Weiss, Rosner, and Sparrow, 2002). Furthermore, optimistic individuals are more likely to recover from illness faster than less optimistic individuals (Sörensen, Chapman, Duberstein, Pinquart, and Lyness, 2017; Achat, Kawachi, Spiro, DeMolles, and Sparrow, 2000). These studies strongly suggest that elderly individuals should work on developing a more optimistic character and enjoy more health benefits during later adulthood. It is quite known that people with an optimistic character have better health and a stronger immune system and have more successful relationships with others. There are many ways to train yourself to be more optimistic. Some people confuse the word pessimism with being "realistic"; they think that predicting negative outcomes in later adulthood is being realistic and something expected. Pessimists may indicate that if something negative happened to others, it will happen to them as well, hence predicting negative outcomes. However, it is important not to try to predict the future based on what has happened to others. Instead, select another role model that experienced positive outcomes. Another way to enhance your optimistic tendency is to spend more time with positive people who help you appreciate the good in your life and spend less time with those who perceive aging as a negative outcome. Spending time with negative people who continually see the bad in every situation will make you feel negative too. On the contrary, spending time with a happy spouse, friend, or neighbor increases the probability that you will be happy as well.

Similarly, the self-esteem factor of psychological resources has shown a strong positive impact on coping with stressful events and enhancing psychological well-being (Luo and Waite, 2011). Self-esteem refers to a person's generally positive evaluation of the self in terms of "worth" and "competence" (Cast & Burke, 2002). Individuals who reported having high levels of self-esteem reported fewer depressive symptoms than those with low levels of self-esteem during late adulthood. In addition, high self-esteem was associated with more respect from other people. Therefore, it is very important to increase the awareness of elderly people and people around them about the impact of self-esteem on health during later adulthood.

Personal control describes people who are able to control or influence their life outcomes. People who perceive that they have personal control are reported to have better health outcomes (Agrigoroaei, Neupert, and Lachman 2013). Perceived personal control is positively associated with important outcomes related to health and well-being, including enhanced cognitive functioning. Naturally, there are individual differences in terms of perceptions of personal control depending on their physiological, behavioral, and emotional health. However, the important message of this chapter is that elderly individuals must learn how to have control over their lives so that they can enjoy better life satisfaction and better health in general. When you believe you have control over how you go about, you will be able to overcome stressors more easily than individuals who perceive that control is externally dependent. Since perceived control is closely associated with health conditions, it is important to maintain health so you can have control over your health outcomes. When you enhance your perceived control, the chances are high that you will have better life satisfaction and well-being.

The other component of psychosocial resources, namely **social support,** refers to the belief of having people in your life who care for you and will help you when you need it (Wills, 1991). The majority of researchers have consistently reported that social support is closely associated with positive psychological well-being and helps in coping during stressful events (Taylor, 2011). As mentioned and detailed earlier in this book, social support also contributes to maintaining or enhancing physical health.

This theory of psychosocial resources, including optimism, self-esteem, sense of control, and social support, showed a strong relationship with life satisfaction and health across the lifespan (Lara, Vázquez, Ogallar, and Godoy-Izquierdo, 2020; Luo and Waite, 2011; Taylor, 2011). These studies were confirmed with a large number of elderly individuals. Therefore, greater attention should be directed towards promoting optimism, strengthening self-esteem, personal control, and fostering positive social support among older adults.

Strongest Factors

From the past literature and current studies regarding aging perspectives, we have come to understand that people who have functioned well during middle and younger adulthood will continue to do so in later life as well. On the contrary, if people do not pay attention to their lifestyles during young adulthood, they are more likely to complain during later life than individuals who functioned well, indicating that aging does not lead to depression or any other psychological issues. As stated earlier, the success of aging depends on previous and current personal lifestyles, past experiences, and social circumstances. As expected, the success of aging well depends on many factors, as mentioned in previous chapters. In this section, the author has selected factors that showed the most significant influence on the successful aging of elderly people in Kosovo. Among the strongest factors were levels of mental health, respect, personality, and religiosity. Therefore, it is important that younger and older individuals themselves, and others (family, relatives, and friends) interacting with elderly individuals, contribute to improving the concerns identified during the research. These results were derived from qualitative and quantitative analysis.

Mental health refers to how we feel, think, and behave, and it affects our overall well-being. It can be characterized by whether we experience any disturbances in our thoughts, emotions, or behavior. Research has shown that mental health is closely linked to depression, which is a condition characterized by persistent sadness and loss of interest in activities. One possible explanation for the strong connection between mental health and depression is that negative life events, such as the loss of friends, spouses, relatives, or declining physical health, are more common among elderly individuals compared to younger people. These negative events can have a significant impact on their mental well-being, potentially leading to the development of depression. This explanation is supported by a study conducted by Kraaji et al. in 2002, which found a correlation between negative events and depression. Furthermore, another study by Njegovana et al. in 2001 discovered a strong association between declining mental health and an increase in major symptoms of depression. This suggests that a decline in mental well-being can contribute to the development or worsening of depression. Additionally, mental health plays a crucial role in influencing a person's overall psychological well-being. Fakouri and Lyon (2005) proposed that experiencing mental disturbances can have a negative impact on an individual's sense of well-being. Considering that an increase in depression can lead to even more critical health issues, it is advisable for elderly individuals to adopt a more constructive approach toward the changes and challenges they experience. It is essential to recognize that the aging process varies from one person to another, as highlighted by Zarit and Zarit (1998). Some individuals may experience a decline in memory relatively early in life, while others may exhibit little to no decline at all. Aging involves a complex interplay of abilities, where certain aspects may decline while others improve. This indicates that aging does not follow a universal pattern for everyone. While cognitive aging studies often focus on decline, it is important to consider that new qualities, such as wisdom, can emerge with age, as proposed by Baltes (1997). As individuals grow older, they become more aware of both the meaningful and meaningless aspects of life. Aging does not imply disengagement from life but rather a different way of engaging with it, driven by different motives and a more focused mind.

"Aging is not lost youth but a new stage of opportunity and strength." – Betty Friedan

Respect refers to appreciation and recognition of elderly people's opinions and thoughts. People in later life expect more consideration; the more their ideas are admired, the more respected they feel. Respect was found to be another strong predictor of stress (Hoxha, 2019). This result indicates that the level of respect given to elderly people in Kosovo determines their levels of stress. No previous literature was found with respect to the relationship between respect and stress. However, the Kosova elderly people in this sample reported that the loss of respect makes them feel worthless and stressed. Most of the participants in the study of Hoxha (2019) reported that the key factor to improve the respect towards elderly individuals is the role of parents, grandparents, and other relatives. This emphasizes the crucial role of the family in instilling values at early ages toward elderly people.

"Those who respect the elderly pave their own road toward success." – African proverb.

The findings indicate that there is a strong positive association between respect and life satisfaction, indicating that elderly people are more satisfied with their lives if they receive enough respect from others. In other words, respect in later life plays a very important role in determining the level of life satisfaction. This finding is consistent with a previous study that suggested that the way older people are treated by the younger generation as well as peers can have a significant impact on their reports of life satisfaction and psychological well-being (Powell, 2005).

"A youth that does not cultivate friendship with the elderly is like a tree without roots." – Ntomba Proverb.

As mentioned in earlier chapters, elderly people in Kosovo have openly stated that the quality of respect from family and loved ones decreased with aging. It was a special request from elderly people about the decisions made in their families. They did not intend to reject suggested decisions made by their close family; they simply wanted to be consulted or at least acknowledged in advance about the major decisions made within or about the family. Decisions like buying, selling, traveling, house innovations, etc. Respect had a particularly strong impact on the elderly participants in Kosovo because of past tradition; the current elderly people did not make final decisions during their young adulthood; they consulted their fathers or grandfathers. Therefore, they reported being disappointed when they are not consulted by their immediate family members.

Personality was reported to have a strong influence on the levels of psychological well-being, which was also supported by Isaacowitz and Smith (2003), who reported that personality is a strong predictor of the aging experience; the more extroverted the person, the more positive experiences of aging reported. The result indicated that personality and perception of aging predict the levels of life satisfaction. Personality and perception of aging are interrelated. The personality type of elderly people influences the perception of aging, and then perception of aging determines the level of life satisfaction.

During the interview with elderly people, the researcher intended to understand the reasons why elderly people were more likely to attribute negative events to aging rather than lifelong patterns of aging. As observed in previous studies as well, elderly people tended to confuse characteristics of aging with disease. For instance, people of certain personality characteristics attributed the negative changes that occur in dementia to **aging,** not disease (Zarit and Zarit, 1998). People who possess characteristics of this type of behavior towards aging do not generally become more pleasant as they age. Because they attributed the negative expectations about aging to old age itself, rather than to a lifelong pattern of aging. They were not aware of the potential lifelong factors that could have affected their attitude towards negative expectations of aging.

Elderly individuals who scored better in extroversion type of personality reported more positive experiences of aging, which in turn was associated with higher levels of life satisfaction and psychological well-being. The way people interpreted the process of aging played the major role in being satisfied or unsatisfied, and this interpretation generally depended on the type of personality characteristics of elderly people. As mentioned earlier, people who were better in traits like openness, agreeableness, consciousness, and extroversion interpreted aging more positively than those who scored low in those traits. Furthermore, it is important to understand that longitudinal studies have found that the type of personality during young adulthood is correlated with personality type later in life (Zarit and Zarit, 1998).

Hence, it can be posited that perceptions regarding aging experiences exhibit consistency across the lifespan, implying that if younger adults perceive aging as a negative phenomenon, it is probable for them to maintain a negative perspective on aging in later stages of life. Similarly, the American Psychiatric Association has documented that individuals encountering challenges in adapting to personality traits during early adulthood are prone to encountering similar difficulties in subsequent stages of life. This suggests that personality is not intrinsically linked to the aging process but rather intertwined with the individual's lifelong trajectory. Thus, it is highly advisable to seek professional assistance and support from experts or individuals with relevant experience.

Finally, **religiosity** was also found to be a strong predictor of psychological well-being, which is supported by the study by Kirby et al. (2004), who suggested that spiritual beliefs influence the psychological well-being of elderly people. The findings indicated that there are significant relationships between religiosity and symptoms of depression, stress, and psychological well-being. Some of the questions that were addressed to elderly people were "do you feel that religion is an important source to cope with day-to-day life difficulties," "do you feel that religion gives you spiritual support?" and "do you feel more at peace (calmer) after religious activities," the majority of the elderly people in Kosovo responded with the answer "very much." The findings of the study indicate that religious activities play a significant role in enhancing the aging process among elderly individuals. For many seniors, religious activities serve as a valuable source of comfort and support. This can be attributed to both the spiritual aspects of engaging in religious practices and the social interactions that occur before and after prayer times. In Kosovo, where the population is predominantly Muslim, retired individuals have the opportunity to gather in mosques at least five times a day for prayer. These regular meetings and prayers have been observed to help elderly people avoid isolation and instead foster integration with others in their community. By participating in these religious activities, seniors have the chance to engage in social interactions, establish connections, and strengthen their sense of belonging. For some elderly individuals, the primary benefit derived from religious activities is spiritual support. Engaging in prayers and religious rituals can provide a sense of solace, peace, and purpose, which contributes to their overall well-being. The spiritual aspect of religious activities offers a source of comfort and guidance, aiding elderly individuals in coping with the challenges and uncertainties associated with aging. On the other hand, social support also plays a crucial role in the positive aging process of elderly individuals. The social gatherings that take place before and after prayer times in mosques provide opportunities for seniors to connect with others, share experiences, and build relationships. These interactions contribute to reducing feelings of loneliness and isolation that can often be prevalent among the elderly population. It is important to note that whether individuals find spiritual or social support through religious activities, both options contribute to successful aging. The combination of spiritual and social elements creates a holistic approach to well-being for elderly individuals. By finding solace in their spiritual beliefs and engaging in social interactions within their religious community, elderly individuals can experience improved overall quality of life, increased social integration, and a sense of purpose, ultimately contributing to successful aging.

Conclusion

Based on past literature and recent studies, it has been observed that successful aging is not solely determined by age but is influenced by various factors. Individuals who have led fulfilling lives during their younger and middle adulthood are more likely to continue functioning well in later life. On the other hand, those who have not paid attention to their lifestyle choices during young adulthood are more prone to experiencing difficulties in later life.

It is important to emphasize that aging itself does not lead to depression or other psychological issues. The success of aging is dependent on a combination of factors, including personal lifestyle choices, past experiences, and social circumstances. In the context of Kosovo, several factors have been identified as significant predictors of successful aging among elderly individuals.

Mental health plays a crucial role in the overall well-being of older adults. It is a determining factor in the prevalence of depression among the elderly population. Negative life events such as the loss of loved ones, declining physical health, and social isolation can impact the mental health of older individuals. Decline in mental health is strongly associated with an increase in symptoms of depression. Therefore, it is important for elderly individuals to maintain a constructive outlook and adapt to the changes and challenges they face.

Respect towards the elderly is another vital aspect that affects their well-being. Elderly individuals in Kosovo expressed a desire for more consideration and respect from their families and society. The level of respect they receive significantly impacts their stress levels and overall life satisfaction. The role of family, parents, grandparents, and other relatives in instilling values of respect during early ages is crucial for fostering a positive environment for older individuals.

Personality traits have a strong influence on the psychological well-being of older adults. Studies have shown that extroverted individuals tend to have more positive experiences of aging and report higher levels of life satisfaction. Perception of aging is intertwined with personality traits, and individuals with traits like openness, agreeableness, conscientiousness, and extroversion interpret aging in a more positive light. It is important to note that personality traits established during young adulthood tend to remain consistent throughout life.

Religiosity has been found to be a significant predictor of psychological well-being among the elderly. Spiritual beliefs and engagement in religious activities can positively influence mental and emotional well-being. Religious practices provide comfort, spiritual support, and social interactions, contributing to successful aging. In the context of Kosovo, where the majority of the population is Muslim, elderly individuals find solace and community through regular mosque gatherings and prayer.

Overall, the success of aging is a multifaceted outcome influenced by various factors such as mental health, respect, personality traits, and religiosity. It is essential for both younger and older individuals, as well as their family, relatives, and friends, to contribute to improving these factors identified through research. By understanding and addressing these concerns, it is possible to promote successful aging and enhance the well-being of older adults. Seeking support from experts or experienced individuals is also advisable for those facing challenges related to personality traits or adjusting to different life stages.

References

Achat H, Kawachi I, Spiro A, 3rd, DeMolles DA, Sparrow D. (2002).Optimism and depression as predictors of physical and mental health functioning: The normative aging study. Annals of Behavioral Medicine, 22, 127–130.

Agrigoroaei, S., Neupert, S. D. and Lachman, M. E. (2013). Maintaining a sense of control in the context of cognitive challenge: Greater stability in control beliefs benefits working memory. GeroPsych (Bern);26:45–49.

Arens, D. A. (1982). Widowhood and well-being: An examination of sex differences within a causal model. International Journal of Aging and Human Development, 15, 27-40.

Atchley, R. C. (2000). Social Forces and Aging. Belmont, CA: Wadsworth.

Berg, A. I., Hassing, L .B, McClearn, G. E. and Johansson, B. (2006) What matters for life satisfaction in the oldest-old? Aging & Mental Health 10(3): 257-264.

Bonanno, G., Wortman, C., Lehman, D., Tweed, R., Haring, M., Sonnega, J. et al. (2002). Resilience to loss and chronic grief: A prospective study from preloss to 18-months post loss. *Journal of Personality and Social Psychology*, 83, 1150–1164

Beaumont, J. G., & Kenealy, P. M. (2004). Quality of life perceptions and social comparisons in healthy old age. Aging and Society, 24, 755–769.

Bailis, D. & Chipperfield, J. (2002) Compensating for losses in perceived personal control over health: A role for collective self-esteem in healthy aging. The Journals of Gerontology, 57b, 6, 531.

Baltes, P. B. (1997). On the incomplete architecture of human ontogeny: Selection, optimization, and compensation as foundation of developmental theory. American Psychologist, 52, 366–380.

Baltes, M. M. & Carstensen, L. L. (1996) Social networks in adult life and a preliminary examination of the convoy model. The Journals of Gerontology, 42, 519-527.

Bandura (1977, 1993) in Fry, P S. (2001) Predictors of health-related quality of life perspectives, self-esteem, and life satisfactions of older adults following spousal loss: An 18-month follow-up study of widows and widowers. The Gerontologist, 41, 6, 787.

Barrow, G. M. (1996) Aging, the Individual, and Society. New York, West Publishing Company.

Bayer, T., Tadd, W. & Krajcik, S. (2005a) Quality in Ageing. Health & Medical, 6, 1 22

Bayer, T., Tadd, W. & Krajcik, S. (2005b) Dignity: The voice of older people. Quality in Ageing, 22.

Belsky, J. K. (1990) The Psychology of Aging. California, Brooks/Cole.

Blanchard-Field, F., Stein, R. & Watson, T. L. (2004) Age differences in emotion-regulation strategies in handling everyday problems. The Journals of Gerontology, 59b, 6, 261-269.

Bradburn, N. M. (1969) The structure of psychological well-being. In Robinson, J. R., Shaver, P. R. & Wrightsman, L. S. (1991) (Eds) Measures of Personality and Social Psychological Attitudes, San Diego: Academic Press.

Braunstein, D. (2004) State strategies to promote independence among older residents.
http://www.srcaging.org/transportation.html. Accessed on 10/1/2007

Brink, T. L., Yesavage, J, A., Lum, O., Heerseman, P. H., Adey, M. & Rose T. L. (1982). Screening test for Geriatric Depression. In Robinson, J. R., Shaver, P. R. & Wrightsman,

L. S. (1991) (Eds) Measures of Personality and Social Psychological Attitudes, San Diego: Academic Press.

Cagney, K. & Lauderdale, D. S. (2002) Education, wealth, and cognitive function in later life. The Journals of Gerontology, 57b, 2, 163.

Jang, I. H., & Choe, S. J. (2006). Social welfare for older persons in an aging society. Seoul, South Korea: Hakjisa

Cast, A. D., & Burke, P. J. (2002). A theory of self-esteem. *Social Forces*, 80, 1041–1068.

Cheng, S. T. & Chan, A. C. M. (2006). Filial Piety and Psychological Well-Being in Well Older Chinese, The Journals of Gerontology: 61(5), 262–269, https://doi.org/10.1093/geronb/61.5.P262

Connidis, I. (1989). The subjective experience of aging: Correlates of divergent views. Canadian Journal on Aging, 8, 7-18.

Coppin, A., Ferrucci, L., Lauretani, F. & Phillips, C. (2006) Low socioeconomic status and disability in old age: Evidence from the in Chaint study for the mediating role of physiological impairments. The Journals of Gerontology, 61, 1, 86.

Dalton, D., Cruickshanks, K., Klien, B. & Klein, R. (2003) The impact of hearing loss on quality of life in older adults. The Gerontologist. 43, 5, 661.

Dein, S., & Stygall, J. (1997). Does being religious help or hinder coping with chronic illness? A critical literature review. Palliative Medicine, 11, 291-298.

Depner, Charlene E. & Ingersoll-Dayton, Berit. (1988) Supportive relationship in later life. Psychology and Aging, 3, 4, 177-180

Diwan, S., Jonnalagadda, S. & Balaswamy, S. (2004) Resources prediction positive and negative affect during the experience of stress: A study of older Asian Indian immigrants in the United States. The Gerontologist, 44, 5, 605.

Diener, E., & Fujita, F. (1997). Social comparisons and subjective well-being. In B. P. Buunk & F. X. Gibbons (Eds.), Health, coping, and well-being: Perspectives from social comparison theory (pp. 329–357). Lawrence Erlbaum Associates Publishers.

Dysktra, P. A. (1995) Loneliness among the never and formerly married: The importance of supportive friendship and desire for independence. The Journals of Gerontology, 50, 321.

Edison, C., Boardman, J., Williams, D. & Jackson, J. (2001) Religious involvement, stress, and mental health: Findings from the 1995 Detroit area study. Social Forces. 80, 1, 251.

Everard, K., Lach, H., Fisher, E. & Baum, M C. (2000) Relationship of activity and social support to the functional health of older adults. The Journals of Gerontology. 4, 208.

Fakouri, C. & Lyon, B. (2005) Perceived health and life satisfaction among older adults: The effects of worry and personal variables. Journal of Gerontological Nursing Thorofare, 31, 10, 17-24

Featherstone, M and Wernick, A. (1995). Images of Aging. London: Routledge.

Featherstone, M., & Hepworth, M. (1991). The mask of aging and the postmodern life course. In Featherstone, M and Wernick, A. (1995). Images of Aging. London: Routledge.

Fennell, G., Phillipson, Ch. and Evers, H. (1988) The Sociology of Old Age. England: Open University Press.

Ferraro, K. F. (1984). Widowhood and social participation in later life. Research on Aging, 6, 451-468.

Festinger, L. (1954). A theory of social comparison processes. Human Relations, 7, 117–140.

Frieswijk, Buunk, Steverink and Slaet (2004).The Effect of Social Comparison Information on the Life Satisfaction of Frail Older Persons. Psychology and Aging, Vol. 19, No. 1, 183–190 Fiori, K. L., Autonucci, T. C. & Cortina, K. (2006) Social network

typologies and mental health among older adults. The Journals of Gerontology, 61b, 1, 25.

Fry, P S. (2001) Predictors of health-related quality of life perspectives, self-esteem, and life satisfactions of older adults following spousal loss: An 18-month follow-up study of widows and widowers. The Gerontologist, 41, 6, 787.

Frey BS, Stutzer A (2002) Happiness and Economics. Princeton University Press: Princeton and Oxford.

Gana, K., Alaphilippe, D., & Bailly, N. (2004). Positive illusions and mental and physical health in later life. Aging & Mental Health, 8, 58–64.

George E. Vaillant, M.D., and Kenneth Mukamal, M.D. (2001) Successful Aging. American Journal of Psychiatry. 158, 6, 839-847.

Gwozdz W, Sousa-Poza A (2010) Aging, Health and Life Satisfaction of the Oldest Old: An Analysis for Germany. Social Indicators Research 97(3): 397–417.

George, L.K. (2010). Still happy after all these years: Research frontiers on subjective well-being in later life. *The Journals of Gerontology: Series B, Volume 65B, Issue 3, Pages 331–339.*

Gerstorf, D., Hoppmann, C. A., Löckenhoff, C. E., Infurna, F. J., Schupp, J., Wagner, G. G., & Ram, N. (2016). Terminal decline in well-being: The role of social orientation. Psychology and Aging, 31, 149–165

George and Bearon (1980) in Krause, N. (2004) Lifetime trauma, emotional support, and life satisfaction among older adults. The Gerontologist, 44, 5, 615.

Hatch, L. R. (2005) Gender and ageism. Generations San Francisco, 29, 3, 19-24.

Havighurst (1986) in Quadagon, J. (1999) Aging and the Life Course: An Introduction to Social Gerontology. USA, McGraw-Hill.

Hayer, W. J., Rybash, J. M. & Roodin, P. M. (1999) Adult Development and Aging. New York: McGraw-Hill.
Hawkley, L., Berntson, G., Engeland, Ch., Marucha, Ph., Masi, Ch. & Cacioppo, J. (2005) Stress, aging, and resilience: Can accrued wear and tear be slowed? Social Science Journals, 2, 115.

House, J, S., Lepkowski, J. M., Kinney, A. M., Mero, R. P., Kessler, R.C. and Herzog, A. R. (1994). The social stratification of aging and health. Journal of Health and Social Behavior, 35, 213-234.

House, J. S., Lantz, P. & Herd, P. (2005) Continuity and change in the social stratification of aging and health over the life course. The Journals of Gerontology, 60b, 15.

Hoxha, A. (2019). Psychological Factors Influencing Adjustment to Retirement. Prizren Social Science Journal, 3(3), 22–31. https://doi.org/10.32936/pssj.v3i3.117

Isaacowitz, D. & Smith, J. (2003) Positive and negative effects in very old age. The Journal of Gerontology, 58b, 3, 143.

Jaewon Lee, M.S.W, and Allen, J. (2021). Connections between Successful Aging and Mental Health among Older Adults: The Role of Artistic Leisure Activities. Journal of Depression and Anxiety Disorders. 3, 1, 80-81.

Jik Joen Lee (2005) An exploratory study on the quality of life of older Chinese people living alone in Hong Kong. Social Indicators Research, 71 335–361

Katz, I.R., Lebowitz, B.D., Pearson, J.L., Schneider, L.S., Reynolds, C.F., Alexopoulos G. S., Bruce, M.L., Conwell, Y., Meyers, B.S., Morrison, M.F., Mossey, J., Niederehe, G. & Parmelee, P. (1998). Diagnosis and treatment of depression in late life. Consensus statement update. Journal of the American Medical Association, 278, 14, 1186-90.

Kesici S, Erdogan A. Mathematics anxiety according to middle school students' achievement motivation and social comparison. Education. 2010;131(1):54-63.

Kutubaeva, R .Zh. (2019). Analysis of life satisfaction of the elderly population on the example of Sweden, Austria and Germany. Population and Economics 3(3): 102-116.

Kubzansky, L. D., Wright, R. J., Cohen, S., Weiss, S., Rosner, B., & Sparrow, D. (2002). Breathing easy: A prospective study of optimism and pulmonary function in the normative aging study. Annals of Behavioral Medicine, 24, 345-353.

Keyes, C. L., B Michalec, R Kobau & H Zahran. (2005) Social support and health-related quality of life among older adults. Morbidity and Mortality Weekly Report, 54, 17, 433.

Kim, J. E., & Moen, P. (2002). Retirement transitions, gender and psychological well-being: a life-course ecological model. Journal of Gerontology: Psychological Science, 57B, 3, 212-222.

Kirby, S., Coleman, P. & Daley, D. (2004) Spirituality and well-being in frail and nonfrail older adults. The Journals of Gerontology 59b, 3, 123.

Kraaij, V., Arensman, E. & Spinhoven, P. (2002) Negative life events and depression in elderly persons: A meta-analysis. TheJournals of Gerontology, 57b, 1, 87.

Krause, N. & Thompson, E. (1998). Cognitive functioning, stress, and psychological well-being in later life. Journal of Mental Health and Aging, 4, 155-170.

Krause, N. (1991) Stress and isolation from close ties in later life.
Journal of Gerontology, 46, 4, 183.

Krause, N. (1991) Stressful events and life satisfaction among elderly men and women. Journal of Gerontology, 3, 84.

Krause, N. (2002) Church – based social support and mortality. The Journals of Gerontology, 61b, 3, 140.

Krause, N. (2003). Stress, social support, and health in late life: Key issues for future research. Contemporary Gerontology, 10, 3-6.

Krause, N. (2004) Lifetime trauma, emotional support, and life satisfaction among older adults. The Gerontologist, 44, 5, 615.

Krause, N. (2005) Stressor arising in highly valued roles, meaning in life, and the physical health status of older adults. The Journals of Gerontology, 59b, 5, 287.

Lang, F. R., Staudinger, U. M. & Carstensen, L. L. (1998) Perspectives on socioemotional selectivity in later life: How personality and social context do (and do not) make difference. The Journals of Gerontology, 53b, 1, 21-39.

Lara R, Vázquez ML, Ogallar A, Godoy-Izquierdo D. (2020).

Psychosocial Resources for Hedonic Balance, Life Satisfaction and Happiness in the Elderly: A Path Analysis. Int J Environ Res Public Health. 17(16):5684. doi: 10.3390/ijerph17165684.

Laukkanen, P., Karppi, P., Heikkinen, E. & Kauppinen, M. (2001) Coping with activities of daily living in different care settings. Age and Aging. 30, 6, 489.

Levinson (1980, 1987) in Stuart-Stuart-Hamilton, I. (1994) The Personality of Aging. London and Bristol, Jessica Kingsley Publishers.

Long, M V. and Martin, P. (2000) Personality, relationship closeness, and loneliness of oldest old adults and their children. The Journals of Gerontology, 311.

Lucas, R., Clark, A., Georgellis, Y., & Diener, E. (2003). Reexamining adaptation and the set point model of happiness: Reactions to changes in marital status. Journal of Personality and Social Psychology, 84, 527–539 Luo, Y., & Waite, L.J. (2011). Mistreatment and psychological well-being among older adults: exploring the role of psychosocial resources and deficits. *Journal of Gerontology: Social Sciences,* 66(2), 217–229.

Luntz, S. (2000) Better off dead. Australian Science, 21, 4, 15.

Lyyra, T. & Heikkinen R. (2006) Perceived social support and mortality in older people. The Journals of Gerontology, 61b, 3, 147.

Martin, M., Grunenthal, M. & Martin, P. (2001) Age differences in stress, social resources, and well-being in middle and older age. The Journals of Gerontology, 56b, 4, 214.

McNeely, R. L. (1988, October). Age and job satisfaction in human service employment. Paper presented at the 39th Annual Scientific meeting of the Gerontological Society of America, Chicago.

McCann, R., Dailey, R., Giles, H. & Ota, H. (2005) Beliefs about intergenerational communication across the lifespan: Middle age and the roles of age stereotyping and respect norms. Communication Studies, 56, 4, 293.

Moos, R., Schutte, K., Brennan, P. & Moos, B. (2005) The interplay between life stressors and depressive symptoms among older adults. The Journals of Gerontology, 60b, 4, 199.

Neugarten, B. L., Havighurts, R. J. & Tobin, S. (1961) The measurement of life satisfaction. In Robinson, J. R., Shaver, P. R. & Wrightsman, L. S. (1991) (Eds) Measures of Personality and Social Psychological Attitudes, San Diego: Academic Press.

Njegovan, V., Man-Son-Hing, M., Mithchell, S. & Molnar, F. (2001) The hierarchy of functional loss associated with cognitive decline in older persons. The Journals of Gerontology, 56A, 10, 638.

Noelker, I. S., & Harel, Z. (2000). Linking quality of long-term care and quality of life. New York: Springer.

Oxman, Th. E., Hull, J. G. (2001) Social support and treatment response in older depressed primary care patients. The Journals of Gerontology, 56B, 1, 35.

Pargament, K. L. (1997) The Psychology of Religion and Coping. New York, Guilford Press.

Parsons (1942) in Featherstone, M and Wernick, A. (1995). Images of Aging. London: Routledge.

Perez, F. R., G. Fernandez-Mayoralas, F. E. P. Rivera and J. M. R. (2001) Aging in place: predictors of the residential satisfaction of elderly. Social Indicators Research, 54, 173–208.

Powell, S. (2005) The World Assembled on Aging, http//www.stridemagazine.com/articles/.
Accessed on 23/6/2006.

Prus, S. & Gee, E. (2003) Gender differences in the influence of economic, lifestyle, and psychosocial factors on later-life health. Canadian Journal of Public Health, 94, 4, 306.

Quadagon, J. (1999) Aging and the Life Course; an Introduction to Social Gerontology. USA, McGraw-Hill.
Robinson, J. R., Shaver, P. R. & Wrightsman, L. S. (1991) (Eds) Measures of Personality and Social Psychological Attitudes, San Diego: Academic Press.

Robinson, S. A. and Lachman, M. E. (2017). Perceived Control and Aging: A Mini Review and Directions for Future Research. Gerontology. 63(5): 435–442.

Rogers, D. (1986) An Introduction to Aging. New York: Prentice Hall.
Rowe and Kahn (1987) in Strawbridge, W., Wallhagen, M. & Cohen, R. (2002) Successful aging and well-being: Self-rated

compared with Rowe and Kahn. The Gerontologist, 42, 6, 727.

Sarah, H. and Pat, Th. (1989) The consolidation of "old age" as a phase of life, 1945-1965, in Featherstone, M and Wernick, A. (1995). Images of Aging. London: Routledge.

Seaman, T. & Chen, X. (2002) Risk and protective factors for physical functioning in older adults with and without chronic conditions: MacArthur studies of successful aging. The Journals of Gerontology, 57b, 3, 135.

Shinkawa, M., Yamaya M., Ohrui, T., Arai H. & Sasaki H. (2002) Depression in older people. Geriatrics and Gerontology International, 2, 215-216.

Shimidt, N. & Sermat, V. (1983) Measuring loneliness in different relationships. In Robinson, J. R., Shaver, P. R. & Wrightsman, L. S. (1991) (Eds) Measures of Personality and Social Psychological Attitudes, San Diego: Academic Press.

Shmidt, R.M (1994). Healthy aging into the 21st century. Contemporary Gerontology. 1, 3-6. Sörensen, S., Chapman, B. P., Duberstein, P. R., Pinquart, M., & Lyness, J. M. (2017). Assessing future care preparation in late life: Two short measures. Psychological Assessment, 29(12), 1480–1495. doi:10.1037/pas0000446

Sperazza, L. (2001) Rehabilitation options for patients with low vision.

The Journal of Rehabilitation Nursing, 26, 4, 148.

Steverink, N., Westerhof, J. G., Bode, Ch. & Dittmann-Kohli, F. (2001) The personal experience of aging, individual resources, and subjective well-being. The Journals of Gerontology, 56b, 6, 364.

Strawbridge, W., Wallhagen, M. & Cohen, R. (2002) Successful aging and well-being: Self-rated compared with Rowe and Kahn. The Gerontologist, 42, 6, 727.

Strydom, H. (2005) Perceptions and attitudes towards adding in two culturally driven groups of aged males: A South African experience. Aging Male. 8, 2, 81.

Stuart-Stuart-Hamilton, I. (1994) The Personality of Aging. London and Bristol, Jessica Kingsley Publishers.

Sung, K. T. (2004). A cross-cultural study of elder respect: Exploration of forms practiced by Americans and Koreans. Journal of Aging Studies, 18, 215–230.

Sung, K. T. (2007). Respect and care for the elderly: The East Asian Way. Lanham, MD: University Press of America.

Taylor, S. E. (2011). How psychosocial resources enhance health and well-being. In S. I. Donaldson, M. Csikszentmihalyi, & J. Nakamura (Eds.), Applied positive psychology: Improving everyday life, health, schools, work, and society (pp. 65–77). Routledge/Taylor & Francis Group.

Thomas, G., Williams, M., Christianna, S., Richardson, E. D. & Tinetti, M. E. (1996). Impairments in physical performance and cognitive status as predisposing factors for Functional dependence among nondisabled older persons. The Journals of Gerontology. 1, 283.

Tomas, J. M., Sancho, P., Gutierrez, M., & Galiana, L. (2014).

Predicting Life Satisfaction in the Oldest-Old: A Moderator Effects Study. Springer Science+Business Media Dordrecht, 1-14 DOI 10.1007/s11205-013-0357-0 Turner, R. J., Lloyd, D. A. & Taylor, J. (2006) Physical disability and mental health: An epidemiology of psychiatric and substance disorders. Rehabilitation Psychology, 51, 3.

Uskul, A. K. & Greenglass, E. (2005) Anxiety, stress, and coping. psychological well-being in Turkish – Canadian sample. Stress Anxiety and Coping. 18, 3, 269 – 278

Van Solinge, H. (2006). Adjustment to and satisfaction with retirement.

Journal of Gerontology: Social Sciences. 60, 1, 11. Vitman, A., Iecovich, E. and Alfasi, N. (2014). Ageism and Social Integration of Older Adults in Their Neighborhoods in Israel, The Gerontologist, 54(2), 177–189.

World Health Organization (1984) Constitution of the World Health Organization. Geneva Zarit, S. H., & Zarit, J. M. (1998). Mental disorders in older adults: Fundamentals of assessment and treatment. Guilford Press.

Zarit, S.H., Todd, P. A., & Zarit, J.M. (1986). Subjective burden of husbands and wives as caregivers: A longitudinal study. The Gerontologist, 26, 260-266.

Zimmer, Z. & House, J. (2003) Education, income, and functional limitation transitions among American adults: contrasting onset and progression. International Journal of Epidemiology. 32, 6, 1098.

www.ingramcontent.com/pod-product-compliance
Lightning Source LLC
Chambersburg PA
CBHW031427210526
45464CB00005B/2080